AF327368

Fertility and Chromosome Pairing: Recent Studies in Plants and Animals

Editor

Christopher B. Gillies, B.Agr.Sc., M.Agr.Sc., Ph.D.

School of Biological Sciences
University of Sydney
Sydney, New South Wales, Australia

CRC Press, Inc.
Boca Raton, Florida

Library of Congress Cataloging-in-Publication Data

Fertility and chromosome pairing : recent studies in plants and
   animals / editor, Christopher B. Gillies.
        p. cm.
      Includes bibliographies and index.
      ISBN 0-8493-6039-0
      1. Fertility. 2. Chromosomes. 3. Meiosis. I. Gillies,
Christopher B., 1943-
QH485.F47 1989
574.3′2--dc19   88-37168
                                                                                                      CIP

Direct all inquiries to CRC Press, Inc., 2000 Corporate Blvd., N.W., Boca Raton, Florida, 33431.

© 1989 by CRC Press, Inc.

International Standard Book Number 0-8493-6039-0

Library of Congress Card Number 88-37168
Printed in the United States

PREFACE

This volume brings together a few selected reviews of recent work on chromosome pairing, particularly electron-microscopic studies of synaptonemal complex formation and behavior, and discusses how pairing may affect fertility. Recent advances in fields as disparate as human *in vitro* fertilization and plant genetic engineering have brought about a realization that many aspects of the control of fertility are poorly understood. The mass propagation of most genetically engineered plants, for example, will continue to rely on sexual reproduction, so that high fertility will be a necessary requirement for their commercial success. The bringing together in genetically engineered organisms of genetic material from taxonomically diverse origins, may have the potential to create unexpected fertility problems.

The increasing trend in most countries towards legal controls on human embryo manipulation, means that more precise information about causes of infertility is essential at the stage of deciding which couples are most likely to benefit as participants in an *in vitro* fertilization program. Such information will also allow better advice to be given to infertile couples in their quest to discover alternative methods of circumventing partial infertility. Conversely, the same information may be useful in designing appropriate programs for contraception or limiting fertility. In the field of conservation, a knowledge of cytogenetic controls on fertility could be of great importance in maintaining or increasing small remnant populations of endangered species. At this point, an understanding of what factors are important for adequate meiotic chromosome pairing and disjunction could well be vital in such cases.

Among the numerous factors which can influence fertility in both animals and plants, the chromosome pairing and crossing over events of meiotic prophase I (from leptotene to pachytene) are crucial to the production of genetically balanced haploid gametes. Early studies of meiotic chromosome behavior by such people as Darlington and McClintock led to a realization of the nature of prophase I pairing, the concept of homology, and the role of chromosome pairing in genetic crossing over. Pioneering cytogenetic studies by Belling, Burnham, Kihara, McClintock, and others showed that structural or numerical changes to the chromosome constitution of an organism resulted in disturbances to chromosome pairing which were often reflected in effects on gametogenesis and fertility. The identification over 30 years ago by Moses and Fawcett of the synaptonemal complex as the basis of meiotic chromosome pairing, opened the way for electron-microscopic analysis of both structural and functional aspects of pairing, and how they might impinge on fertility.

After the initial descriptive phase, progress has been slow in characterizing the structural and functional aspects of the synaptonemal complex. The recent application of monoclonal antibody and molecular biological techniques now promises to allow the identification of the components of the synaptonemal complex. Most of the details of the mechanism of meiotic chromosome pairing, whereby the lateral elements of homologous leptotene chromosomes come together to form a synaptonemal complex which allows reciprocal crossover events to occur, have remained unresolved problems. The questions being asked do not differ in substance from those which light-microscope cytogeneticists had speculated on prior to the description of the synaptonemal complex.

Early ultrastructural studies using random thin sections revealed that it was not always possible to define clearly the basis of homologous pairing; infertile plants and animals which were asynaptic at metaphase I might often appear to have normal synaptonemal complexes at pachytene. The introduction of serial-sectioning techniques, together with three-dimensional reconstruction of complete lateral elements and synaptonemal complexes, confirmed that at pachytene in a diploid organism synaptonemal complexes stretched from end to end of each fully paired bivalent. Serial sectioning also allowed the investigation of questions about the dynamics of synaptonemal complex behavior, although the sample sizes investigated were often insufficient for clear answers on all aspects. Such studies have shown that in the majority of cases, chromosome pairing commences at or near the telomeres of chromosomes, which are usually attached to a limited area on the inner side of the nuclear envelope. Studies in organisms with abnormal chromosome constitutions, such as structural rearrangements and numerical variants, revealed that nonhomologous pairing could occur and often resulted in apparently normal-looking synaptonemal complexes. With the recognition by Carpenter that the structures she named "recombination nodules" were another part of the machinery of pachytene crossing over, synaptonemal complex formation was identified as a necessary but not entirely sufficient prerequisite for meiotic crossing over.

The last 10 years have seen the development of a number of spreading techniques which have allowed the routine analysis of synaptonemal formation at meiotic stages from leptotene to diplotene in both plants and animals. While it is true that spreading techniques, by their nature, result in the loss of information about three-dimensional aspects of synaptonemal-complex behavior, they have been used for both light- and electron-microscopic studies of a wide variety of organisms and chromosomal variants, and have allowed the accumulation of significant numbers of results in many cases. In this volume, the authors have drawn on these results to illustrate the recent advances in the field and how they may help in our understanding of problems of fertility.

The first two chapters summarize the wealth of data from studies in humans and mice, with information on meiotic and synaptonemal complex behavior, and the effects of numerical and structural changes in both autosomes and sex chromosomes. In their reviews, both Speed, and de Boer and de Jong emphasize the importance of prophase I pairing saturation in determining fertility, and show that heterologous pairing can occur as a primary event and not just by synaptic adjustment. The data from humans and mice show that there are often differences in meiotic chromosome pairing behavior in males and females, and emphasize the rarity of such comparisons in most other organisms. The interaction of incompletely paired autosome segments (such as from structurally rearranged and aneuploid chromosomes) with sex chromosomes, also emerges as an important aspect of prophase I chromosome behavior which can be a determinant of fertility in mice and humans.

In his chapter, Solari describes the pairing behavior of sex chromosomes in the heterogametic sex of both mammals and birds, particularly concentrating on the nature and extent of this pairing in different sex chromosome systems, and its influence on fertility. In this field, the recent advances in the use of DNA probes have allowed the confirmation of the pseudo-autosomal behavior of the human X and Y chromosomes, and brought into focus the question of the significance of synaptonemal complex formation and crossing over between the X and Y. The authors of these chapters agree that biochemical studies with monoclonal antibodies and immunolabeled probes will be important in extending investigations beyond the strictly descriptive stage to the analysis of functional aspects of synaptonemal complex formation in animals. The pioneering biochemical work, particularly in lily microsporocytes, of Herbert Stern's group has led the way in this field.

Plant chromosomes are in general much larger than those of most animals, and studies of the ultrastructure of their pairing are consequently more complicated. Serial sectioning and spreading studies are only just beginning to produce a meaningful body of data about plant pairing behavior. Jenkins has summarized the situation in plant hybrids, illustrating several well-documented examples of prophase I pairing in interspecific hybrids. Once again, one of the striking results is the occurrence of extensive heterologous pairing, including length adjustment of chromosomes of widely different DNA contents. The existence of genetic control of pairing, particularly the ability to discriminate between different levels of homology, exists in hybrids, but becomes of greater significance in polyploids. As shown in previous chapters, such controls appear to be widespread, and are found in quite a few polyploids, including many autopolyploids. The ability to maximize pairing appears to extend in some polyploids to a drive to even-numbered pairing configurations, which obviously should lead to more balanced gametes and higher fertility.

In spite of this, the widespread occurrence of triple pairing, noted in particular by Speed, de Boer and de Jong, and Gillies, indicates that saturation of pairing can be achieved by another means. It remains to be seen what the significance of triple pairing is with regard to subsequent crossing over and chromosome disjunction. It does, however, illustrate that a drive to maximize or saturate pairing appears to be a factor in prophase I chromosome behavior, but that it can be achieved in a variety of ways. If pairing saturation is an important factor in determining the successful completion of gametogenesis, as Burgoyne has postulated, then there are a number of ways to achieve it. The data presented here certainly support the idea, and reiterate the importance of chromosome pairing for fertility.

In tribute to his pioneering ultrastructural work on meiotic chromosome pairing, I would like to dedicate this volume to Professor Montrose Moses. Not only was he the co-discoverer of the synaptonemal complex in 1956, but again in the 1970s he was at the forefront of the development of the spreading techniques which have been the basis of much of the most recent research described in this volume. He has continued to contribute to the field with work on substaging of pachytene, and on biochemical aspects of synaptonemal complex function.

C. B. Gillies

THE EDITOR

Christopher B. Gillies, Ph.D., is a Senior Lecturer in Genetics, School of Biological Sciences, University of Sydney, New South Wales, Australia.

Dr. Gillies graduated in 1965 from the University of Queensland, St. Lucia, Australia, with a B.Agr.Sc. degree, was awarded First Class Honors in Plant Breeding and Genetics in 1966, and a M.Agr.Sc. degree in 1968. In 1970, he graduated with a Ph.D. in Genetics from the University of Alberta, Edmonton, Canada.

Dr. Gillies spent from 1970 to 1973 as a postdoctoral fellow in the Institute of Genetics at the University of Copenhagen, Denmark, and then 1 year as an Amanuensis in the Department of Physiology at the Carlsberg Laboratory Copenhagen. He has been at the University of Sydney since 1974.

Dr. Gillies is a member of the Genetics Societies of Canada, America, and Australia, currently being Honorary Secretary of the latter society. He has presented papers, posters, and invited lectures at national and international meetings in Genetics and Cell Biology, and has been a guest speaker at universities in the U.S., Canada, and several European countries. His research work has been published in more than 30 research papers, several major reviews, and conference proceedings, and in about 20 conference abstracts. His current research interests include the study of synaptonemal complexes in humans, animals, and plants, with the aim of investigating aspects of meiotic chromosome-pairing behavior and their bearing on fertility.

CONTRIBUTORS

P. de Boer, M.Sc., Ph.D.
Department of Genetics
Wageningen Agricultural University
Wageningen, The Netherlands

J. H. de Jong, M.Sc., Ph.D.
Department of Genetics
Wageningen Agricultural University
Wageningen, The Netherlands

C. B. Gillies, B.Agr.Sc., M.Agr.Sc., Ph.D.
Senior Lecturer
School of Biological Sciences
University of Sydney
Sydney, NSW, Australia

G. Jenkins, B.A. (Hons), Ph.D.
Department of Agricultural Sciences
University College of Wales
Aberystwyth, Wales

A. J. Solari, M.D.
Full Professor
Centro de Investigaciones en
 Reproducción
Facultad de Medicina
Buenos Aires, Argentina

R. M. Speed, B.Sc., M.Sc.
Chief Research Officer
Human Genetics Units
Medical Research Council
Edinburgh, Scotland

TABLE OF CONTENTS

Chapter 1

HETEROLOGOUS PAIRING AND FERTILITY IN HUMANS

R. M. Speed

TABLE OF CONTENTS

"In all this work we have to be aware that meiosis works with chromosomes which always have the two functions of accomplishing evolution and of implementing its results in heredity. In consequence, the adaption of meiosis is perpetually imperfect."[1]

I. INTRODUCTION

Fertility in mammals, of which humans are perhaps an unique example, has been of interest since primitive times. On a global scale, it would appear that man has been highly successful in this capacity, with a doubling of the world population expected between the present time and the first decade of the next century. Indeed, until recently, the main problem has been the control of fertility. With technological advances in both the understanding of the hormonal control of reproduction and in methods of contraception, unwanted fertility has virtually disappeared in the developed countries.[2] As a result of the apparent ease of such fertility control, it is perhaps not surprising that couples now believe that there are difficulties in fertility if a child cannot be conceived within a short period of time.

On an individual basis, about one in ten couples will encounter problems of conception, with varying levels of accompanying emotional trauma.[3] In 60 to 80% of such couples, factors in either the male or female partner will be the cause of the fertility problem. A combination of events will account for the remainder.

In the context of this chapter, the problems of fertility in humans will be limited to an examination of various aspects of the meiotic process. This is a vital stage in the system by which normal and/or abnormal gametogenesis occurs in humans, leading to the production of male and female germ cells. In particular, the study of the pairing process between homologous human chromosomes at the prophase stage of meiosis will be examined for any subsequent effects on fertility.

Historically, the investigation of the human male germ cell system has predominated due to the more ready availability of testicular material from adult males attending infertility clinics. Since Ford and Hamerton in 1956[4] employed a squash technique to establish that male spermatocytes contained 23 bivalents, subsequent development of air-drying methods[5] has allowed advances in the analysis of homolog synapsis and disjunction at the light-microscopic (LM) level. More recently, the modification of the Counce and Meyer[6] surface-spreading technique to human material has enabled detailed analysis with both the light[7] and electron microscopes[8] of meiotic pairing at the synaptonemal complex (SC) level. Complementary observations on serial sections of human germ cells[9-11] have also provided much detailed information on SC pairing at the electron-microscopic (EM) level in both sexes. The structure of the SC in organisms as diverse as fungi and man is remarkably conserved and its relationship with the meiotic process is well established.[12,13] The origins of the SC at the earliest stages of meiotic prophase and its absence in situations where crossing over is not evident, as in *Drosophila* males, [14] suggest its involvement in both homolog synapsis and recombinational events, both being of critical importance to eventual germ cell development in humans.

II. HOMOLOGOUS VS. HETEROLOGOUS PAIRING IN HUMANS

In reality, very little experimental meiotic work of any nature has been performed on humans due to the obvious ethical problems. The majority of studies of chromosome pairing, the structure and function of the SC, and their relationship to germ cell maturation and fertility have occurred in plant and animal species. Data from rodent[15] and primate[16] sources will, however, be of major interest in defining general concepts and models applicable to humans. Chromosomally normal or abnormal situations, either created experimentally or occurring naturally in such animal systems, are open to extensive investigation and will be described in detail in other chapters. Nevertheless, a brief description of pairing at the chromosome/SC level is appropriate at this point to clarify the nature of heterologous pairing.

Chromosome pairing begins in most species with the rough alignment (300 nm apart) of the lateral elements of the homologous chromosomes.[17] Only when lateral elements are separated by approximately 100 nm does the SC begin to form. Initiation points may be numerous, as in plant species, or be almost strictly telomeric, as in the human oocyte. Normally, only homologs will initiate synapsis at meiotic prophase (homologous pairing), but numerous examples now exist of SCs that can form between chromosomes or segments of chromosomes that are nonhomologous in genetic content (heterologous pairing).[18-20] Heterologous pairing may also develop at prophase, in chromosomal rearrangements such as duplications or inversions,[21] and has been termed by Moses et al.[22] "synaptic adjustment". The fact that such SCs appear of normal dimensions and structure suggests that the SC, per se, is not the mechanism of genetic exchange, as such heterologous pairing in plant haploid species,[23] does not lead to chiasma formation or crossing over.

It is generally held that recombination will only take place when DNA sequences of strict homology are brought into register either in the central region of the SC or within the bulk of the chromatin surrounding the SC. More than 99% of the chromosomal DNA remains outside the confines of the SC, it being calculated for *Neurospora*[12] that the total nuclear SC length of 50 μm represents only 0.3% of the total DNA double helix length of 16 mm. Gillies[24] has shown that the corresponding figure for maize is only 0.014% of the total DNA length. Experimental evidence, however, has favored the proposal that recombination occurs within the SC, primarily because Westergaard and von Wettstein[25] have observed that chiasma appear to cross remnants of the diplotene SC in *Neottiella*. Moreover, Moses and Poorman[26] have found an apparent association between P-DNA synthesis and the central regions of the SC in mouse pachytene oocytes. Such observations are thought to represent evidence of a breakage and reunion mechanism associated with crossing over during the pachytene stage of meiosis.

Alternatively, the initial events of meiotic homolog synapsis may only involve the matching of isolated chromatin blocks or DNA sequences at scattered points along the chromosomes. It is in this way that Riley and Flavell[27] have suggested that chromosomes which are similar only in a gross manner and not in fine structure, could at least synapse, if not progress to recombination. This would be even more likely if the limited number of recognition sites included repeated sequences, which would be of a more similar nature even in nonhomologs. The various models for synapsis involving special sites or sequences have recently been reviewed in detail by Chandley,[27] who has further suggested that early replicating R-band sites could serve as the initiation centers for homologous and/or nonhomologous meiotic pairing.

It may be, however, that heterologous SC synapsis better serves other functions, perhaps related to fertility, than to those ultimately leading to recombination in germ cells. In proposing a theory relating chromosome pairing and fertility in species as varied as *Drosophila* and humans, Miklos[29] noted a correlation between the reduction or absence of synapsis and gametogenic breakdown. It was suggested that a saturation of pairing sites between homologs of either sex chromosomes or autosomes was vital for the normal progression of germ cell development. This was later extended to include both male and female germ cells by Burgoyne and Baker.[30] Many such examples of the operation of this theory have been reported, and Burgoyne and Biddle[31] suggested that the loss of spermatocytes in XYY mice and human males was due to the univalence of the X and Y chromosomes. In human males with either a ring 21[32] or ring E group chromosome,[33] a strong correlation between the failure to pair of a single autosomal bivalent and male sterility has been reported. If heterologous synapsis, as opposed to homologous synapsis, was permissible in situations where pairing failure of either segments or whole chromosomes occurred, would the fertility of the organism be influenced in any way? In the case of human and mouse XO females,[20] it appears that in the former where the single X chromosome fails to synapse in any way, sterility is the usual endpoint. In XO female mice where the single X chromosome undergoes heterologous pairing either with itself or with other autosomes, fertility, though reduced in comparison with XX female mice, is the norm.[20] Again, in the case of human inversion carriers where heterologous synapsis can occur across the inversion region, the higher the level of such heterologous pairing, the more normal the fertility of the patient appears (See later).

III. NORMAL MALE AND FEMALE MEIOSIS

A. The Timing of Meiotic Progression

It is important to realize with regard to human fertility that male and female meiosis, while being processes of similar function, i.e., the production of haploid gametes, are of a very different temporal nature. Initially, the germ cells in both male and female arise in or near the yolk sac endoderm, from where they migrate to areas overlying the mesonephros, which develop into the fetal gonads. From this point on, each sex takes its own separate course of maturation. After proliferating for a number of mitotic divisions, the germ cells in the male invade the tubules of the embryonic testis and become resting gonocytes. This state then lasts until puberty, when meiotic activity commences and the gonocytes progressively become type A and B spermatogonia, 1 and 2° spermatocytes, and, finally, spermatids.

In the human female, however, the germ cells or oogonia within the fetal ovary continue their development, increasing greatly in number by mitotic division. In the human female, approximately 7 million oocytes are present at midterm, but decline in number to about 2 million by birth.[34] The elimination of so many oocytes by degeneration or atresia seems quite a drastic selection when we consider that only 400 to 500 oocytes will finally be ovulated. After oogonia enter the final mitotic interphase, they proceed through a premeiotic DNA synthesis stage and enter the prophase of meiosis as early as week 11 of gestation.[35] By birth, they will have entered a resting stage known as dictyotene, which is characterized by highly diffuse chromosomes. The important feature of human female meiosis, then, is that synapsis and recombination will have taken place by birth. Although the dictyotene stage has been termed a resting stage, just after birth, when the oocyte and the follicle that it is enclosed in are growing in size, active replication of RNA occurs, the chromosomes now having a lampbrush-like structure.[36] Such RNA may be laid down for the subsequent stages of meiosis and early stages of embryogenesis. Only later, as the granulous cells of the follicle become active, supplying maternal proteins to the ooplasm, can the dictyate nucleus be said to be in the resting state. This stage persists until puberty in the female, at approximately 12 years of age. Follicular maturation then takes place and, with cyclical ovulation, a certain percentage of oocytes are deemed mature enough to respond to pituitary gonadotrophin. Some follicles then respond to follicle stimulating hormone and undergo final maturation, and are induced to move out of the dictyate stage and progress through diplotene and metaphase I under the influence of luteinizing hormones (LH). This LH peak is about 36 h before ovulation in the human female. The first meiotic division occurs with the formation of the metaphase II chromosomes and the first polar body. This is the stage where meiotic nondisjunction can first occur, due to the failure of bivalent separation or random segregation of univalent chromosomes. There may be total nondisjunction leading to diploid metaphase II oocytes which, on fertilization, would yield triploid fetuses. The secondary oocyte then arrests at metaphase II and it is at this point that ovulation takes place. No further development occurs unless fertilization takes place, when the second meiotic division occurs with the separation of chromatids and second polar body extrusion. This is the second stage where meiotic nondisjunction can occur.

Table 1
CHROMOSOME ABNORMALITIES FOUND WITHIN A GROUP OF 2372
UNSELECTED MALES ATTENDING AN EDINBURGH SUBFERTILITY CLINIC

Chromosome analysis	Number
47XXY	24
47XYY	5
46XY/47XYY	1
45X/48XYYY	1
45X/46,r(Y)	1
46X,inv(Y)(pll;qll)	1
47XY,mar+	4
Robertsonian translocation	4
Reciprocal autosomal translocation	10
Total	51

From the above, several important features stand out. First, spermatogenesis is an ongoing process, with multiplication from basic stem cells, whereas the number of oocytes in the human ovary is fixed at birth. Second, meiosis in the male is completed in approximately 74 d,[37] while in the human female it is spread over an exceedingly long period of time, varying from 10 to 40 years or more.[36] Third, with reference to potential fertility resulting from male and female meiosis, the extrusion of two polar bodies in the female allows for the elimination of chromosomally abnormal complements since the polar bodies eventually degenerate. This is not possible in the male, where all four products of meiosis are initially viable sperm.

B. Human Meiosis at the Light-Microscopic Level
1. Male

As previously stated, the analysis of human male meiosis has predominated over the last 20 years. Only a few studies of truly normal, healthy, fertile individuals have been reported,[38] the majority of testicular material having been obtained from patients attending subfertility clinics. From the latter groups it became apparent, as data were gathered, that males carrying chromosomal abnormalities occurred with a frequency much above that of the general population. A figure of 5.3% of chromosomally abnormal males was obtained from the combined data of the four largest surveys undertaken,[33,39-41] which is approximately three times the rate in the newborn population. It also became evident that as the sperm count was lowered, the number of chromosomal abnormalities substantially increased. From the Edinburgh subfertility survey[39] comprising 2375 males, for a sperm count in the range of 21 to 60 $\times$ 10^6/ml, only 0.94% abnormalities were recorded, but with a sperm count reduced to 1 $\times$ 10^6/ml or less abnormalities rose to a high of 15.38%. The composition of chromosomal abnormalities found at Edinburgh is given in Table 1. It is clear that sex chromosome abnormalities are strongly represented, particularly the 47XXY class, these falling within the severe oligospermic to azoospermic range. Translocations of both the Robertsonian and reciprocal type were recorded, as they have been consistently so in other surveys. The nature and effects of these chromosomal abnormalities on subfertility, as seen at the LM level, have been extremely well reviewed elsewhere[42] and only more recent evidence from EM studies will subsequently be dealt with. The great majority of the testicular material obtained from subfertility clinics has been from males with a normal karyotype. From this material, the stages of prophase from early zygotene to metaphase II have been clearly defined.[43] Centromere position within bivalents has been determined by modification of the ''C'' banding technique[44] and chiasma position has been mapped using fluorescent staining techniques.[45] At the pachytene stage, details of the isolation of the X and Y chromosomes within the sex vesicle have been investigated.[46] It has also been recognized that the sex chromosomes at meiotic prophase are condensed, late labeling, and show early cessation of transcription relative to the autosomes.[47] Studies of chromomere maps have been developed using the pachytene stage. Bivalents appropriately treated[48,49] exhibit dark and light staining regions of chromatin which show close approximation of size and number with the "G" banding patterns seen in the somatic chromosomes of mitotically dividing cells.

2. Female

The major problem in the study of human female meiosis arises because of the separation of the initial prophase stages within the fetal ovary from the subsequent meiotic divisions in the adult, both of which have necessitated the development of specialized techniques. Initially, simple squash techniques[50-52] allowed indentification of the prophase stages with the LM and it was determined that they appeared toward the end of the first trimester of fetal development. Further clarification of pairing was provided by the air-drying method of Luciani and Stahl,[53] chromomere patterns allowing the identification of individual bivalents. The study of the first and second metaphase chromosomes has relied on *in vitro* cultures of mature oocytes released from the follicles of the adult ovary. This technique has been successfully used for mammalian species as varied as mouse[54] and monkey[55] and has provided data ranging from the effectiveness of pairing and disjunction to chiasma frequencies. Further application of the technique to man by Edwards[56] has highlighted the difficulties of working with human female meiosis. Because of the ethical difficulties of human superovulation, few oocytes become available for experimentation in this manner. Ovarian biopsies were previously obtained at the time of hysterectomy operations, but currently the conservation of the ovaries as functional hormonal organs has precluded this source. Indeed, throughout the scientific literature, very few reports of clear human female metaphase I or II chromosomes exist.[57-61] With the technique, metaphase I chromosomes appear within 24 h, while after 48 h the metaphase II complement are visualized. For the human female, it seems that chiasma frequency is lower in oocytes[43] than in the male spermatocyte, whereas for other mammalian species, where more extensive data are available,[62] the reverse would appear to be true. Currently, research involving human oocytes and early postfertilization embryos is a matter for legislation. To overcome the low 10% success rate of *in vitro* fertilization, research to determine optimum culture and development rates is required. The probable time restriction for research to 14 d post fertilization will hopefully permit the necessary improvements of technique required in this vital area of human subfertility.

Surveys of subfertile adult human females to determine the influence of chromosome abnormalities, already shown to be of importance in male subfertility, have been limited. Of some 850 women at an Edinburgh subfertility clinic, Jacobs[63] found only 5 to be karyotypically abnormal. Three reciprocal and one Robertsonian translocation, plus one extra marker chromosome were identified. The resulting frequency of 0.59% abnormal was not significantly different from the 0.38% obtained among normal female controls.

C. Human Meiosis at the Electron-Microscopic Level

1. Male

The adaptation of the surface-spreading technique to the human male spermatocyte has allowed a detailed investigation at the EM level of the prophase stages of meiosis and, in particular, the behavior of the XY bivalent during pachytene. Initial observations demonstrated that pairing commenced at the telomeres of the short arms of the X and Y chromosomes and through early pachytene extended to almost the centromere of the Y chromosome.[64,65] More extensive studies by Chandley et al.[8] proposed that pairing could extend across the centromere and might progress to include the entire euchromatic segment of the Y chromosome. This was recently confirmed by Sumner and Speed[66] using an immunochemical labeling technique which allowed the kinetochores to be accurately labeled within the XY bivalent, showing that pairing did extend beyond the Y centromere. To what extent such pairing between the human X and Y chromosomes is truly homologous as opposed to heterologous is currently a matter for debate, involving questions as to the origins of the heteromorphic sex-determining chromosomal systems of mammals and as to why they are segregated within the sex vesicle and out of synchrony with the autosomal complement.

It had initially been assumed by Koller and Darlington[67] that at least a segment of homologous pairing occurred between the X and Y chromosomes, within which at least one cross over could occur. This was further supported by Ohno et al.,[68] who suggested that the end-to-end associations seen in many mammalian XY sex chromosome systems resulted from the terminalization of an already nearly terminal chiasma. More recently, Burgoyne[69] has proposed that pairing between the X and Y chromosomes is, to a certain point, a consequence of homology, albeit confined to a small terminal segment. An obligatory cross over is initiated within such a region and genes distal to the chiasma would be inherited in a pseudoautosomal manner. Confirmation of such a theory is, however, difficult, as the mapping of such genes will be extremely complicated on the basis of regular pedigree analysis. The only clear example of an apparently autosomally inherited mutant for which an autosomal location has been ruled out is the sex reversing (Sxr) mutation[70] in mouse. The mutation in XY males has involved the duplication and transference of the testis-determining factor from a proximal location on the Y chromosome to the distal tip beyond the normal pairing region. An apparent exchange event then transfers this Sxr fragment at meiosis to the X chromosome,[71] determining that females who inherit this X chromosome will be phenotypically male. The exchange of chromosomal material has been cytologically confirmed by Evans et al.[72] A consistent feature of metaphase I in such Sxr males is the high proportion of cells with an unpaired X and Y chromosome. Figures of 70 to 90% unpaired, as compared with 5 to 10% in normal controls, have been obtained by several authors.[73,74] A recent study at the EM level has shown that the Y chromosome with two testis-determining sequences has a strong drive to initiate self-pairing and completely synapse heterologously[75] at the expense of regular synapsis with the X chromosome. In this case, the heterologous pairing of the Y chromosome may contribute to the varying fertility in Sxr males since an association between spermatogenetic failure and lack of X and Y chromosome pairing has long been known.[76,77]

Alternatively, Ashley[78] has taken the view that apart from ''end attraction'', synapsis between the X and Y chromosomes is mostly heterologous in nature, suggesting that, of the 200 genes that have been assigned to the human X chromosome,[79] little evidence of similar loci residing on the Y chromosome is evident. Further, Ashley[78] suggests that such heterologous XY pairing, taken in combination with a premature desynapsis of the sex chromosomes during the prophase of meiosis, relative to the autosomes, is the basis of a mechanism that has evolved to maintain regular disjunction, but to prevent crossing over. Evidence relating to the asynchronous synapsis of the XY bivalent with regard to the autosomes is presented in support of such a proposal. While the autosomes are paired throughout pachytene, autoradiography studies have detected a DNA repair system thought to be related to exchange events occurring at mid-pachytene. By this time, however, the sex chromosomes of mouse and man have desynapsed. Again, even though recombination nodules have been identified on the sex bivalent of man[10,65] at early pachytene, when pairing between the X and Y chromosomes is at a maximum, it is the number present at mid-pachytene which is thought to more truly represent the number of chiasmata seen at metaphase I.[80] Recombination nodules are not observed on the XY pair after early pachytene[81] and desynapsis of the X and Y chromosomes occurs at approximately the same time. Mutants such as the Sxr male mouse have to be explained on the basis of an aberrant delay in desynapsis which will permit such exchanges to take place between the X and Y chromosomes.

Recent observations suggest that a combination of the above proposals is permissable. DNA probes have indicated that a region of homology exists between the mammalian sex chromosomes. The MIC2 X and Y DNA sequences are both located at the distal ends of the X and Y short arms,[82] as are the hypervariable telomeric sequences reported by Cooke et al.[83] and Simmler et al.[84] The latter sequences show, as expected by the Burgoyne theory, pseudoautosomal inheritance. This region, however, is small in relationship to the distance over which the X and Y chromosomes can pair, as visualized at the EM level. The testis-determining factor (Tdf) in the human male has been relocated from a pericentromeric region to the short arm of the Y chromosome.[85] Even though this factor appears to be outside the pseudoautosomal region, occasional XX males have been reported whose DNA profiles contain Y-specific sequences. This occurs presumably because of an exchange event between the X and Y chromosomes, as originally proposed by Ferguson-Smith.[86] Rouyer et al.[87] have demonstrated a gradient of crossing over between the human X and Y chromosomes, its frequency decreasing with distance from the telomeres, which might encompass such rare cross overs causing the Tdf exchange. It appears, then, that there is a small telomeric region of pseudoautosomal activity, but that the majority of the X and Y chromosome synapsis is heterologous, preventing a more extensive recombination between the two.

The classification of the XY pairing stages, the progression of synapsis and desynapsis in relationship to the autosomes, has been of importance in the analysis of fertility in the male. Moses et al.[64] and Solari[65] classified six basic substages ranging from Type O at late zygotene, where no apparent SC had yet formed, through to Type V at late pachytene. During this period, the initial homologous SC formation occurs in the Type I nucleus (Figure 1a). Maximum homologous and heterologous synapsis occurs toward the Type I, Type II transition (Figure 1b). In the Type V nuclei, all that remains of the X and Y SC is a tangled mass created by the repeated splitting of the lateral elements. A more detailed description of the XY pairing types has recently been given by Chandley et al.[8] The collection of data from a small group of chromosomally normal males undergoing an orchidectomy operation for adenocarcinoma of the prostate at Edinburgh has allowed a baseline to be established for a normal progression of pachytene spermatocytes, as observed at the EM level (Table 2). This differs a little from the data originally reported by Solari[65] in the classification of Types III, IV, and V pachytene nuclei, but serves to be measured against other observations from patients attending the Edinburgh subfertility clinic (WSM series). Biopsies are obtained which are processed for testicular histology, LM analysis of metaphase I and II chromosomes, and EM analysis of spread pachytene spermatocytes. The sperm count of the WSM patients ranged from 85×10^6 to 0.0×10^6/ml, with a mean of 10.8×10^6/ml. From a total of 1549 cells from 32 patients, the eventual mean distribution of XY pairing types was remarkably similar to that obtained for the control group (Table 2). This would suggest for the majority of patients with a normal karotype that synapsis of the sex chromosomes will not be a major factor contributing to their condition. However, on closer examination, a small group of patients showed variation from the normal range, the mean sperm count $(13.6 \times 10^6$/ml) of the group being at the lower end of the subfertile range. A block as early as late zygotene is apparent in WSM 246, whose sperm count was only 0.5×10^6/ ml. The majority of spermatocytes appeared to be degenerating, with the X and Y in the occasional nuclei attempting to synapse. The X and Y in 50% of the analyzable cells were detached.

The stage most often affected was the Type II pachytene nuclei. Four of the patients had twice the number of spermatocytes at this stage (compared with controls), while WSM 251 appeared to have three times the number of Type III spermatocytes.

Desynapsis of the heterologously paired region, it seems, can occur, but further progression

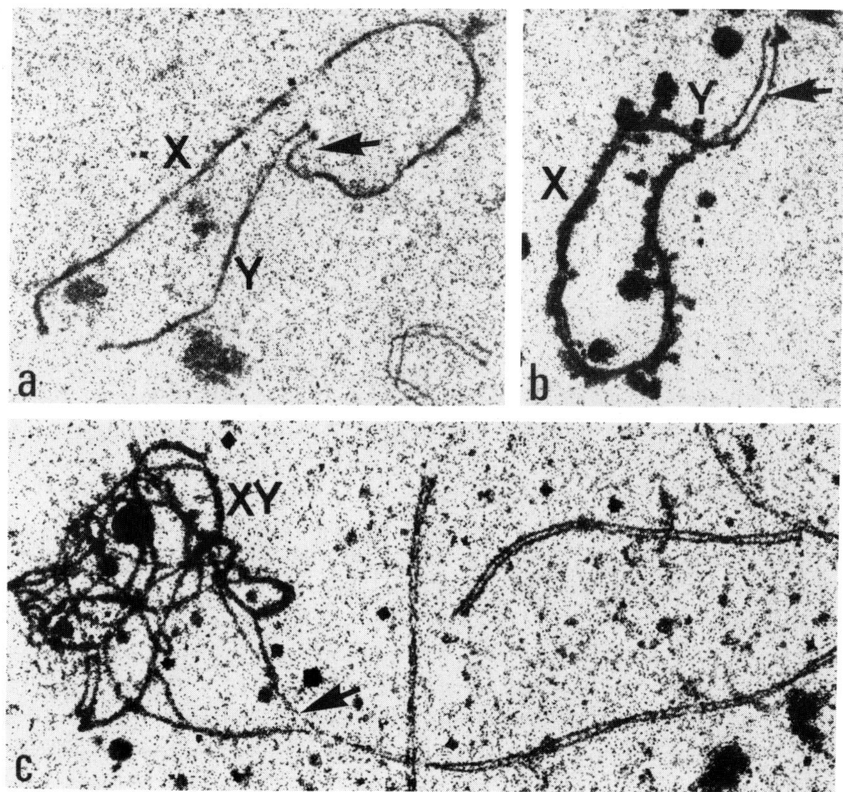

FIGURE 1.   (a) Type I, XY bivalent with only a short SC, including the pseudoautosomal region. (b) Type II, XY bivalent. Almost 50% of the length of the Y chromosome is included in the SC (arrow), most of which represents heterologous synapsis. (c) Autosomal contact with Type IV, XY bivalent (arrow).

Table 2
PERCENTAGE DISTRIBUTION OF SPERMATOCYTES FROM NORMAL AND SUBFERTILE MALES BETWEEN PACHYTENE SUBSTAGES BASED ON XY PAIRING AT THE EM LEVEL

	No. of cells	XY Stage						X and Y unpaired
		0	I	II	III	IV	V	
Normal	531	2.4	10.9	21.1	23.0	34.8	6.6	1.1
Data from *Solari* (1980)	86	2.3	19.8	18.6	5.8	19.8	33.7	Not given
Subfertile								
Chromosomally normal	1549	3.8	8.5	27.2	12.8	35.6	7.0	5.0
Individual patients								
WSM 174	45	4.4	15.6	46.8	8.8	17.7	4.4	2.2
175	50	2.0	2.0	40.0	6.0	38.0	8.0	4.0
181[a]	50	2.0	24.0	42.0	6.0	22.0	2.0	2.0
189[a]	20	0.0	10.0	25.0	30.0	30.0	5.0	0.0
226	50	8.0	12.0	42.0	10.0	18.0	0.0	10.0
246	12	41.7	0.0	8.3	0.0	0.0	0.0	50.0
251	50	2.0	4.0	8.0	64.0	14.0	4.0	2.0
Chromosomally abnormal								
t(9;20)	98	0.0	7.1	27.5	6.1	44.9	13.3	1.0
Inv(I)	72	29.2	23.0	8.3	2.8	2.8	0.0	33.3

[a] XY pairing segment never longer than 15% of the length of the Y chromosome.

to the later stages where the X and Y chromosomes are multiply split is blocked. A further synaptic problem occurred in patients WSM 181 and 189. In addition to a block in progression, the X and Y chromosomes failed initially to pair to the expected extent. The maximum synapsis observed between the X and Y chromosomes never exceeded 15% of the length of the Y chromosome.

Two of the subfertile patients were chromosomally abnormal (Table 2), one carrying a t(9;20) reciprocal translocation, the other an inversion of chromosome 1. Synaptic progression of the XY bivalent in the translocation patient was comparable to the chromosomally normal subfertiles. Other features of chromosomal rearrangement leading to infertility will be dealt with later. The inversion-carrying patient suffered severe problems of XY synapsis. In 30% of the cells examined at the EM level, the sex chromosomes were unattached. A block with very little progression to the Type V nuclei occurred with excess numbers of Types I and II occurring. The detached nature of the sex bivalent will certainly contribute to infertility on the basis of the Miklos theory[29] and problems of synapsis in inversions will be looked at in a later Section.

Little can be suggested as to the meiotic mechanisms operating in the subfertile, but chromosomally normal patients, except that in all cases the autosomal SCs are generally normally synapsed. The timing of the synapsis of the XY within the sex vesicle in relation to the autosomes is obviously abnormal in some cases, but the underlying nature of such events remains to be investigated.

Although the majority of autosomal SCs in the chromosomally normal group of subfertile males synapse regularly, small numbers of aberrant forms are seen. At mid-pachytene, fully paired autosomal SCs appear to degenerate, and multiple nonstaining regions become evident. In degenerating rat oocytes, such cells have been termed Z-cells by Beaumont and Mandl.[88] Further, synaptic errors, including heterologous pairing, occur at low frequency, as shown in Table 3. The differences between the controls and chromosomally normal subfertiles include a threefold increase in the numbers of Z-cells, but with little increase in any of the other categories of synaptic error. A fourfold increase in contact between the XY bivalent and a terminal region of an autosomal SC was observed in the subfertile group, compared with the controls (Figure 1c). In one patient, WSM 234, a 100-fold increase occurred (Table 3). The nature of the XY contact with an autosome is unlikely to represent true synapsis, but a small region of telomeric heterologous pairing cannot be totally ruled out. Contact between the XY bivalent and certain autosomal translocations has been observed by Forejt[89] and its possible influence on fertility will be a subject of later discussion. Of the two chromosomally abnormal patients, the translocation carrier showed disturbance of autosome synapsis. In the inversion carrier, the number of Z-cells was greatly increased and problems of synapsis resulting in autosomal bivalents with asynaptic regions and total failure of pairing, yielding univalents, were common. Such errors, according to the theory of Miklos,[29] would lead to the elimination of such germ cells and an eventual reduction in fertility.

2. Female

An initial approach to the study of human fetal oocytes using the EM was undertaken by Bojko[11] using sectioned cells. This method provides much data on the spatial arrangement of the SCs within the nucleus, but is time-consuming and the number of oocytes that have been examined are extremely limited. As in the male, the surface-spreading technique and the EM have allowed for a far greater detail in analysis of both the progression of the human female germ cells through the prophase stage of meiosis[35] and the process of synapsis. This has allowed us to address problems such as what are the general levels of oocyte degeneration seen in normal fetal ovaries?[90] How do these compare with the even more severe arrest in chromosomally abnormal situations, such as the XO[20] and trisomy 18[91] human fetuses? Also

Table 3

PERCENTAGE DISTRIBUTION OF NORMAL AND ABNORMAL SYNAPTIC TYPES IN HUMAN MALE AND FEMALE PACHYTENE GERM CELLS OBSERVED AT THE EM LEVEL

	Number of:		Percentage								
	Individuals	Cells scored	Normal	Z-cells	Interlocks	Nonhomologous pairing	Triple pairing	Interchange	Asynaptic regions	Univalents	XY contact with autosomes
Normal males	13	549	92.2	4.7	0.2	0.2	0.0	0.0	2.0	0.4	0.4
Subfertile males											
Total chromo-somally normal	32	1578	78.4	13.3	0.2	0.4	0.0	0.1	3.2	0.6	3.8
Individual pa-tient WSM 234	1	50	42.0	10.0	0.0	0.0	0.0	0.0	6.0	0.0	42.0
Chromosomally abnormal											
t(9;20)	1	100	74.0	5.0	0.0	0.0	0.0	0.0	2.0	1.0	18.0
Inv(I)	1	119	37.0	35.3	0.8	0.0	0.0	0.0	16.0	10.9	0.0
Normal females	5	1200	53.6	15.2	1.5	7.6	4.3	2.2	7.1	7.8	N.A.

Note: N.A., not applicable.

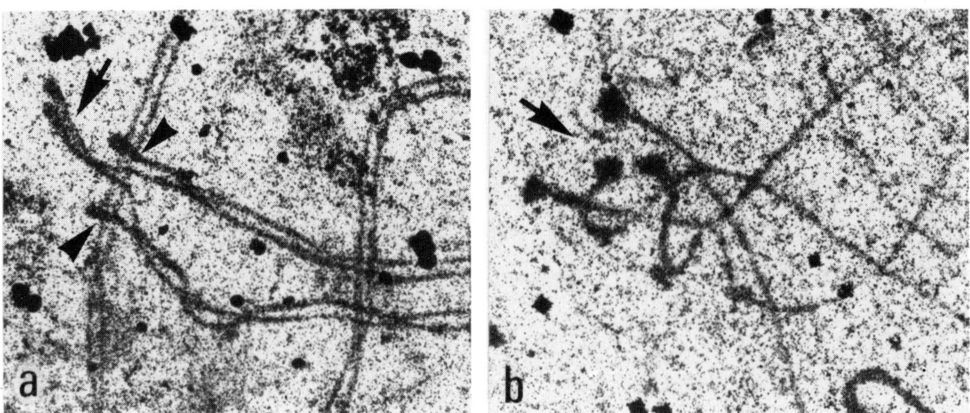

FIGURE 2.   Heterologous pairing in human pachytene oocytes. (a) Almost complete synapsis between pairs of lateral elements of differing lengths (arrow heads) with probable homologous terminal synapsis (arrow) of longest elements. (b) Cluster of four telomeres in close proximity at early zygotene. (From Speed, R. M., *Hum. Genet.*, 78, 260, 1988. With permission.)

open to investigation is the question of whether a production line in the fetal ovaries, as proposed by Henderson and Edwards,[54] can account for the higher rates of nondisjunction seen in mouse and human females of increased maternal age.

Fetal ovaries which are chromosomally normal can be occasionally obtained from abortions carried out for social reasons. They are best obtained within 24 h of extra-amniotic prostaglandin induction. The somatic karyotype is confirmed by fetal blood lymphocyte analysis. Unlike the human male, where the sex bivalent is a convenient marker of pachytene progression, the XX bivalent behaves in a similar manner to the autosomal complement and cannot be distinguished from it. Progression has therefore relied more on nucleolar morphology, length of the SCs, and synaptic and desynaptic behavior.[26] The rates of progression observed by surface spreading in normal human fetal ovaries[35] broadly agree with previous reports from sectioned material[34] and air-dried preparations.[92]

The most striking finding within normal fetal oocytes is the high incidence of synaptic error, in comparison to that found in male spermatocytes (Table 3). Similar observations have been reported in the mouse at the EM level by Mahadevaiah and Mittwoch.[93] Of 1200 human oocytes studied at Edinburgh, only 54% showed a normal synapsis. Z-cells increased threefold over levels in the normal male, more approximating levels seen in the subfertile males. Partial asynapsis within bivalents or complete asynapsis leading to the production of univalents increased 20-fold in oocytes compared with spermatocytes. Heterologous pairing also appears to be a far more frequent feature in the normal female germ cell. The simplest form involves short lengths of lateral elements or univalent axial elements pairing on themselves to form loops or hairpins. Alternatively, two heterologous axes may be found with synapsis occurring over variable distances. Also included in this group are oocytes within which virtually complete heterologous synapsis has occurred. A pair of SCs may exhibit one normally synapsed set of telomeric ends, while the other reflects an asymmetrical pattern, with the longer homologs pairing over a short distance (Figure 2a). Such events could originate at early zygotene when groups of telomeric ends frequently observed in close proximity could undergo mispairing (Figure 2b). A further example of heterologous pairing always involved two SCs which had undergone an exchange of axial elements (Figure 3a,b). Both telomeric ends of the two SCs involved were symmetrically paired, but located a short distance from one set of telomeres was a switch of pairing partners. That these were a translocation-type event can be argued against by the nonrandom subtelomeric position they

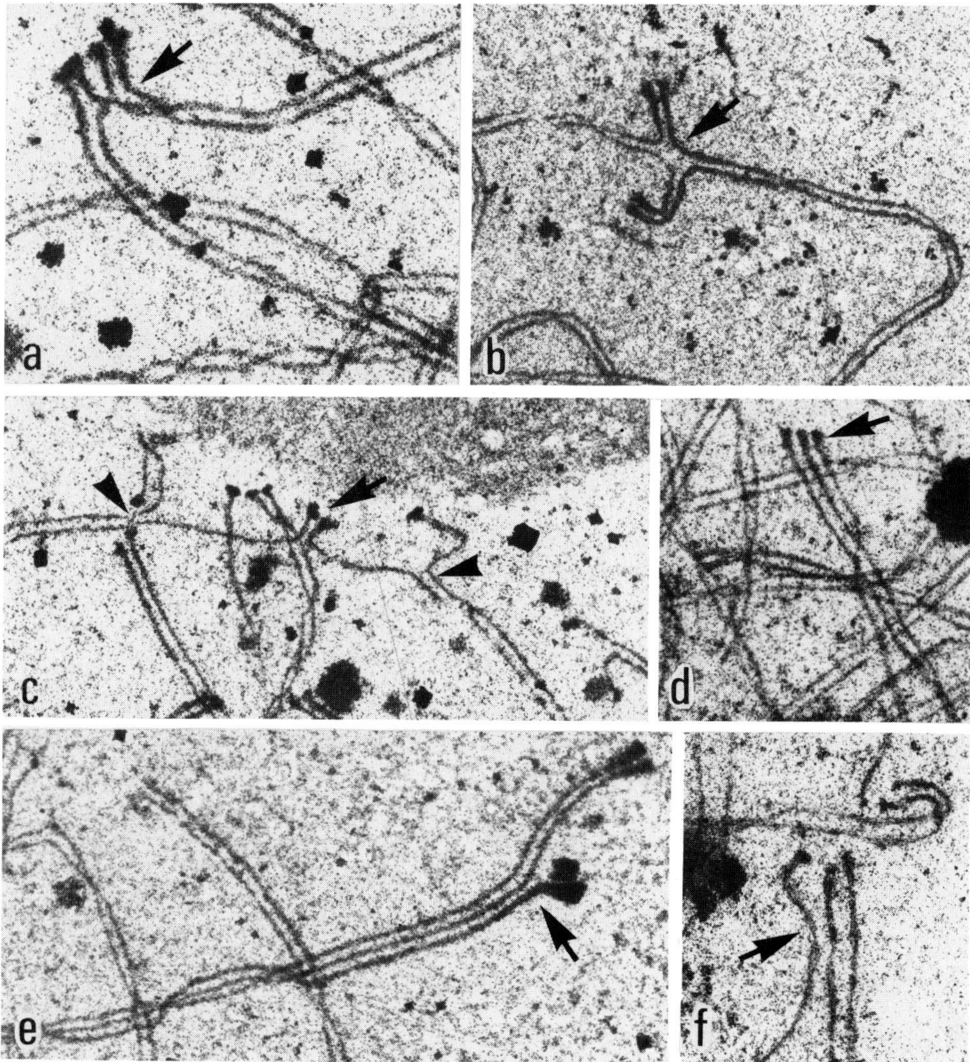

FIGURE 3. Heterologous pairing in human pachytene oocytes. (a,b) Symmetrical exchange of pairing partners in the vicinity of telomeres. (c) Possible origin of exchange event at mid-zygotene. Homologous pairing from subtelomeric regions (arrow heads) with heterologous synapsis from telomeres (arrow). (d) Triple pairing with telomeres in register (arrow). (e) Telomeres of presumed homologs in register with extra lateral element (arrow) asymmetrically and heterologously synapsed. (f) Initiation of triple pairing at zygotene (arrow). (From Speed, R. M., *Hum. Genet.*, 78, 260, 1988. With permission.)

occurred in. Such events appeared to originate at zygotene (Figure 3c). Heterologous telomeres form a small pairing segment, while homologous pairing proceeds from a subtelomeric position. The final form of heterologous synapsis involved the association of three lateral elements. While this might be expected in trisomic situations,[91,94,95] it is unusual in karyotypically normal female fetuses. The three telomeres were most often in register (Figure 3d), while occasionally the associated nonhomolog initiated heterologous synapsis away from the normally paired telomeres (Figure 3e). Again, such a triple synapsis appeared to originate at zygotene (Figure 3f). It has previously been proposed that all initial prophase synapsis is strictly homologous, heterologous synapsis being a secondary event.[96,97] In human fetal oocytes, however, where homologous axial elements failed to pair, heterologous syn-

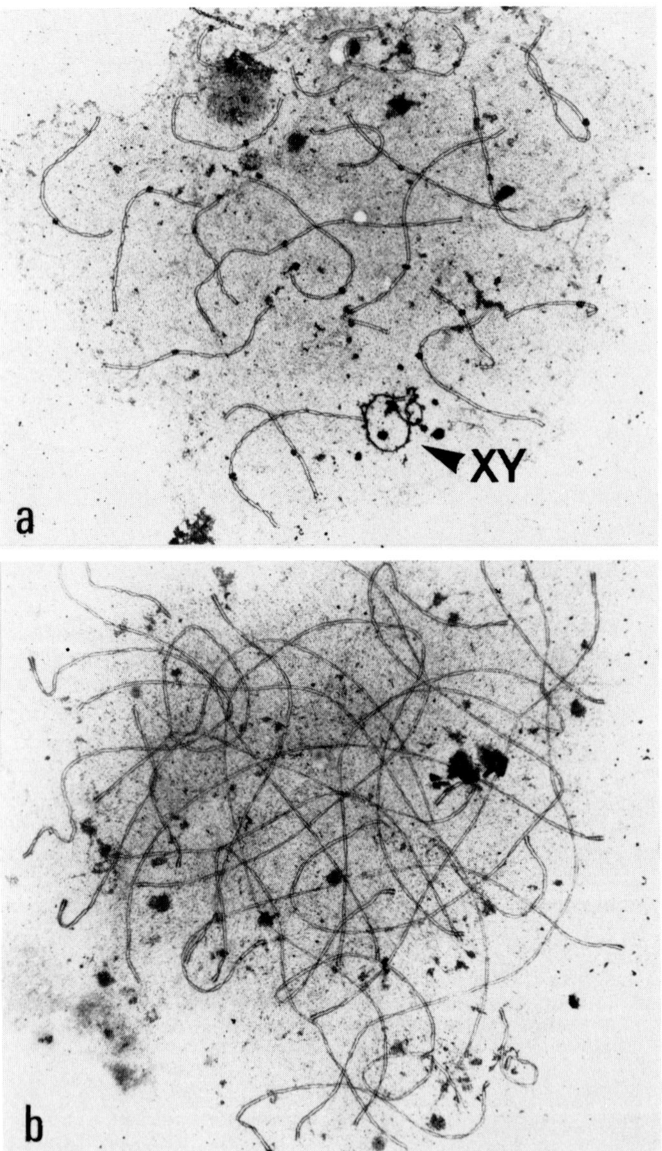

FIGURE 4.   Comparison of SC length in spermatocytes and oocytes. (a) Human pachytene spermatocyte with Type II, XY bivalent (arrow head). (b) Human pachytene oocyte (same magnification as a) with total SC length almost double that of the spermatocyte in a. (From Kew Chromosome Conference Proceedings, No. III. With permission.)

apsis appeared at the same time, usually producing short stretches of SC, but occasionally almost total heterologous synapsis between dissimilar chromosomes was observed. Why female germ cells should be more prone to such synaptic errors is not known. It might relate to the greater overall length of the female SC complement (Figure 4a,b). It has been estimated by Bojko[11] that the SC length in human oocytes is about twice that found in spermatocytes, the nuclei being of similar dimensions. The mechanical constraints of such long SCs may make location of homologous telomeres more difficult in fetal oocytes. Movement to appropriate synapsis points may be delayed, so that regular pairing is prevented, then telomeres may pair heterologously with the nearest available partner in an attempt to comply with

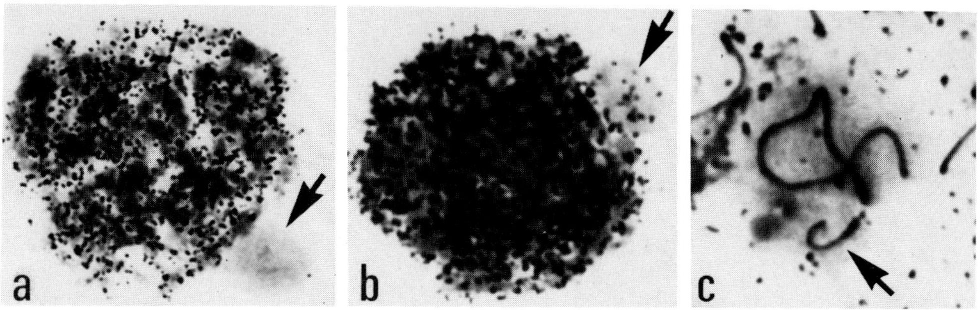

FIGURE 5.   RNA activity in the sex vesicles of chromosomally normal and abnormal male mice. (a) Normal XY male, showing unlabeled sex vesicle (arrow). (b) Tertiary trisomic male mouse, Ts(1^{13})70H, showing labeled sex vesicle (arrow). (c) Surface-spread Ts(1^{13})70H pachytene spermatocyte at LM level. Localized grains in vicinity of 1^{13} marker chromosome (arrow) within the sex vesicle. (From Speed, R. M., *Chromosoma*, 93, 267, 1986. With permission.)

ntranuclear time schedules. The effects on fertility of such synaptic events in the normal human female are difficult to assess. The reduction in numbers of oocytes from 7 million at midterm to 2 million by birth is the norm. The differences of reproductive potential between males and females, whereby an initially small, fixed pool of oocytes and subsequent ovulation of a few mature ova in the adult female will ensure fertility, can perhaps accommodate such a dramatic reduction. The human male, who must produce at least 20×10^6/ ml spermatozoa to be classified as fertile, is liable to be more susceptible to synaptic error.

IV. EFFECTS OF TRISOMIES AND SEX CHROMOSOME ANEUPLOIDY

The only pure autosomal trisomy to survive to adulthood in man, involves chromosome 21, which is the cause of Down's syndrome. A review of the literature by Johannison et al.[98] showed that for the male spermatogenic arrest was common. Sperm counts analyzed by Stearns et al.[99] in nine cases showed four to be azoospermic, the remaining five having low counts in the subfertile range. Johannison et al.[98] have also confirmed earlier meiotic studies and shown that the extra chromosome 21 at metaphase I can be present as an univalent (88.5% of nuclei) or a trivalent (4.8% of nuclei). The remaining 6.7% of the nuclei appeared to have lost the third chromosome 21. Only 8.5% of the surface-spread prophase nuclei, however, contained a clear univalent and 2.4%, an identifiable trivalent. To explain the different observations at pachytene and metaphase I, it was proposed that the extra chromosome 21 was initially included in the sex vesicle with the X and Y chromosomes. The majority of pachytene nuclei analyzed were Types III to V, the tangled morphology of the sex bivalents making the inclusion of extra material difficult to assess. Support for the idea comes from the analysis of tertiary trisomic male mice derived from the T70H translocation.[100] The extra chromosome is almost exclusively contained in the sex vesicle, its thickened appearance resembling that of the sex bivalent. According to the Lifschytz and Lindsley model,[101] where inactivation of the X chromosome is essential to the progress of normal spermatogenesis, any such disruption of the sex vesicle might lead to infertility. In the Ts(1^{13})70H tertiary trisomic male, an attempt to monitor the levels of RNA synthesis[102] in the normally silent pachytene sex vesicle (Figure 5a), showed that aberrant RNA synthesis occurred (Figure 5b). In surface-spread preparations (Figure 5c), it appeared possible that the small marker chromosome retained its normal activity, rather than the X being reactivated. Such a disturbance might, in itself, lead to cell death.

In Down's females, fertility is possible, there being 27 recorded cases of live births. The ratio of normal to trisomy 21 offspring closely approximates the expected 1:1 segregation

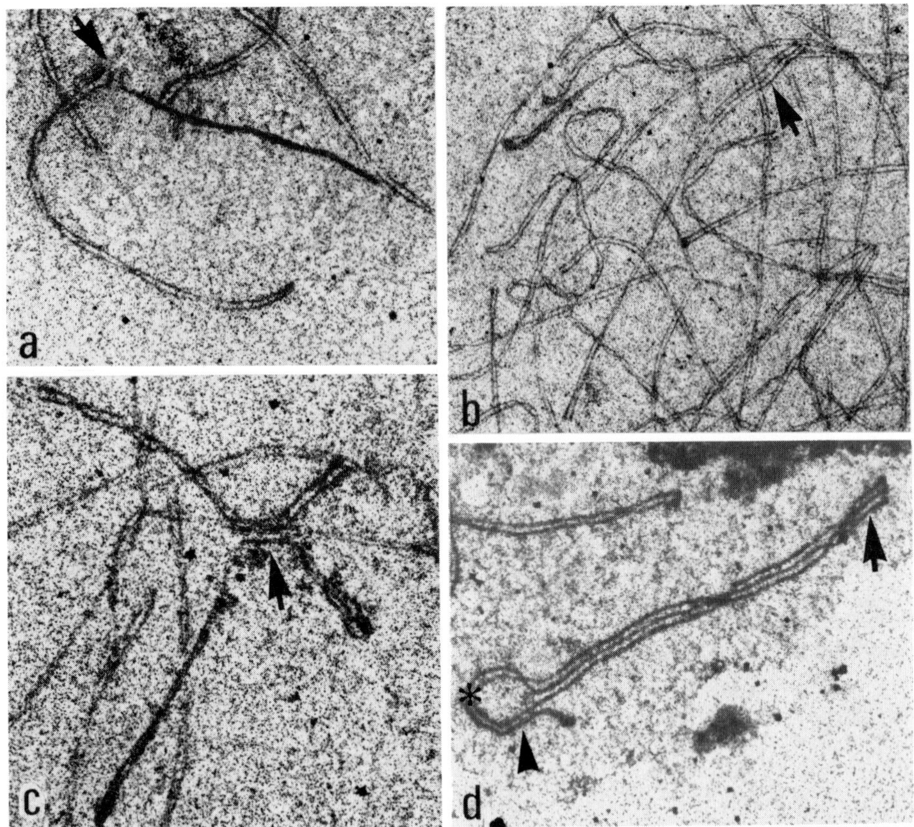

FIGURE 6.   Triple pairing in human aneuploid foetuses. (a) Small terminal pairing segment (arrow) between three chromosome No. 21 lateral elements. (b) Triple synapsis of three No. 21s over one third of their length (arrow). (c) Interstitial triple pairing of three No. 21 chromosomes (arrow). Unpaired regions of lower lateral element are thickened and multiply split. (d) Almost complete triple pairing of a normal chromosome 18 (arrow) with an iso-18q. Homologous synapsis to centromere region (star), with short arm of normal 18, then showing heterologous synapsis (arrowhead). (From Speed, R. M., *Hum. Genet.*, 72, 256, 1986; Speed, R. M., *Hum. Genet.*, 66, 176, 1984. With permission.)

pattern.[103] A study of trisomy 21 fetal oocytes[94] at the EM level has shown that as in the male the three chromosomes 21 can synapse in a triple association (Figure 6a-c). Compared with the male, the trivalent configuration is more frequently observed (33.8% vs. 2.4%), as is the free univalent (60.0% vs. 8.5%). The apparent synaptic success of the third chromosome 21 in the female merely reflects the absence of an isolated, partially asynaptic female sex bivalent. The extra chromosome 21 in the female only has the option to triple pair or remain as an univalent. Increased degeneration of oocytes was not evident in this fetus and if survival to adulthood had occurred, fertility could have been possible. The only other report of trisomy in human fetal oocytes studied at the EM level involved a mosaic 18p-;iso 18 female.[91] Here, the iso 18 cell line was effectively trisomic for the majority of chromosome 18. Triple pairing between the homologous iso 18 arms and the normal chromosome 18 occurred in 68% of the cell line. In the occasional cell, the normal 18 paired both homologously and heterologously with the iso 18 (Figure 6d). A feature of this fetus was an apparent meiotic delay in oocytes reaching the pachytene stage, as compared with chromosomally normal fetuses of the same gestational age. Germ cells appeared to block at a preleptotene stage. Such an effect may account for the severe reduction in germ cell numbers seen in the ovaries of surviving trisomy 18 babies[104] and trisomy 21 girls.[105] In the fetal testes of the trisomy 13, 18, and 21 condition, severe reductions of germ cell numbers also occurs.[106] Delay and degeneration appear then to be characteristic of human aneuploid germ cell development in both males and females, the earliest onset being seen in the fetal gonads.

The most common male sex chromosome aneuploidy is the 47,XXY Klinefelter's syndrome. Such males are generally azoospermic with small testes, lacking any germ cell development. This situation is thought to arise because of the failure of germ cells with two X chromosomes to survive in a testicular environment. In 47,XYY males, fertility is extremely varied, ranging from virtual normality to almost azoospermia.[107] The majority of meiotic data for XYY males suggest that only germ cells of an XY constitution are seen at metaphase I.[108,109] Other studies of sectioned spermatocytes examined at the EM level have argued for the retention of the second Y until the pachytene stage[110] and 45% of spermatocytes in a single XYY patient reported by Hultén and Pearson[111] contained one X and two Y chromosomes at metaphase I. Information from surface spreading is lacking and would face the same problems of analysis as in Down's syndrome males in that the tangled nature of the XY bivalent would make the identification of two Y chromosomes difficult. The variable fertility of such males most likely reflects the presence of normal XY cell lines which would contribute to fertility, the presence of univalents, and a failure to saturate pairing sites in the XYY cell lines, leading to their elimination.

For the human female, the XO chromosomal constitution is the most frequently occurring form of sex chromosome aneuploidy seen (20%) in spontaneous abortion surveys.[42] Those that survive to term and adulthood represent only 5% of the XO conceptions. The condition is not so lethal in the mouse, where two thirds survive to term.[112]

Extremes of fertility are seen between XO human and XO mouse, which may be related to the ability of the single mouse X chromosome to undergo heterologous pairing. Most adult human XO females show a failure of ovarian development, even though fetal ovaries have been shown to contain germ cells.[113] XO mice were initially thought to have normal fertility,[114] but subsequent studies have shown that they too have fewer oocytes than normal and a shortened reproductive lifespan.[115] Burgoyne and Baker[116] suggested that the presence of the univalent X chromosome might be the causal agent of such excess atresia in pachytene oocytes. A recent study of surface-spread fetal oocytes from both XO human and XO mouse[20] has shown differing synaptic capabilities of the single X chromosome in both species. Initially in the human XO fetal ovaries, a block in progression occurs similar to that seen in autosomal trisomies. The majority of oocytes remain in a preleptotene stage. Those few that do progress to pachytene contain a single, thickened axial element interpreted as an X univalent (Figure 7a). In the mouse, a minor delay in progression occurs, but the great majority of oocytes pass through pachytene, reaching the dictyate stage by birth. Here, in a proportion of oocytes, the mouse X chromosome may remain single and thickened, as in the human. The X chromosome in most oocytes, however, undergoes heterologous pairing. This may take the form of self-synapsis, forming hairpins or loops (Figure 7b-d), or, alternatively, each telomere of the X chromosome can initiate heterologous synapsis with two homologous but as yet unpaired autosomal telomeres and synapse to form a triradial structure (Figure 7e). On day 19 of gestation, just before birth, approximately two thirds of the oocytes contained an X chromosome which, although lacking a homologous pairing partner, was able to satisfy its pairing instincts by heterologous synapsis. Survival of oocytes in the XO mouse ovary might, therefore, depend on such pairing behavior and would be in accord with the hypothesis of Miklos,[29] where pairing sites must be saturated to permit a regular maturation of germ cells. The time of initiation of the heterologous synapsis is of interest for it is seen to begin at earliest pachytene. Indeed, when competitive synapsis with an autosome occurs, the X axis is paired along a substantial portion of both autosomal axial elements, suggesting an origin as early as zygotene. This would be contrary to the view that all initial synapsis is

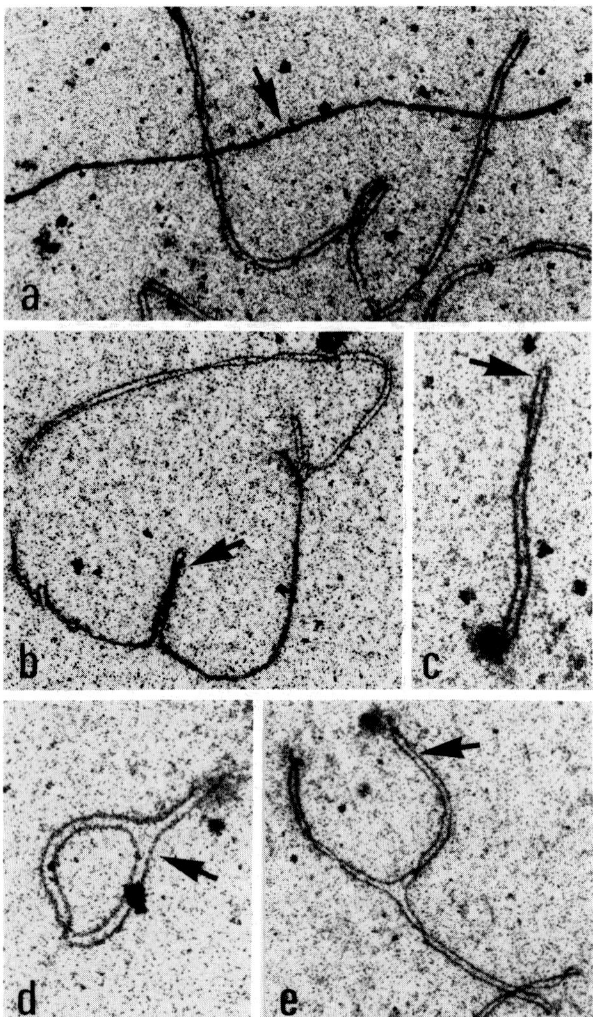

FIGURE 7.   Human and mouse XO pachytene oocytes. (a) Single thickened human X axial element (arrow). (b) Initiation of heterologous pairing in mouse X axial element (arrow). (c) Mouse axial element heterologously synapsed as an hairpin (arrow). (d) Mouse axial element heterologously synapsed as a ring (arrow). (e) Mouse X (arrow) fully heterologously paired with both axial elements of an autosomal SC. (From Speed, R. M., *Chromosoma*, 94, 115, 1986. With permission.)

strictly homologous.[19,96] However, heterologous synapsis as early as zygotene has been reported in other species.[117-119] In the few cases of apparently fertile human XO women, it may be that the single X chromosome can also occasionally undergo heterologous synapsis, allowing oocyte survival into the adult ovary. Further examination of human XO fetal oocytes would be required to confirm such a possibility.

V. TRANSLOCATIONS AND FERTILITY

Spermatogenic breakdown in human male translocation heterozygotes and experimental work, particularly with mouse translocations which contribute to our understanding of germ cell death, have been reviewed by Chandley[42] and Searle.[120] Initially, the interactions of the translocations themselves with the nuclei may produce disturbances and cell death. Secondly, irregular segregation from translocations will create variable numbers of unbalanced gametes, contributing to fetal wastage and an apparent reduced fertility. Problems of synapsis fall within the first category and several hypotheses to explain the relationship of fertility and irregularities of pairing in both human Robertsonian (centric fusion between acrocentric autosomes) and balanced reciprocal translocations (involving combinations of autosomal acrocentrics, nonacrocentrics, and sex chromosomes) will be examined.

A. Autosome-Sex Chromosome Translocations

Both X and Y chromosomes may be involved in human translocations, but the condition is rare. Madan et al.[121] reviewed 14 male carriers of differing X-autosome translocations showing all the adult carriers to be oligo- or azoospermic, germ cell arrest at the primary spermatocyte level being the most common defect. A case involving an X-2 human translocation with associated azoospermia was recently examined at our laboratory using surface spreading.[122] The translocation chromosomes at pachytene formed a chain quadrivalent in 86% of the cells. Even though the X and Y chromosomes were involved in the quadrivalent, the sex bivalent type could still be assessed. Types I and II were mainly observed, indicating an early arrest. Asynapsis was common in the vicinity of the breakpoints. Approximately 18% of the quadrivalents showed heterologous pairing in the central region. This is similar to that reported in two mouse X-17 translocations by Ashley et al.[123] It appears that sections of the X chromosome that lie outside the normal X and Y synaptic region can heterologously pair with lateral elements of an autosomal origin. The fertility of female carriers will depend on where the breakpoints occur. Those located in the "X-critical" region (Xq13-Xq26) will lead to primary or secondary amenorrhoea.[124] True Y autosome translocations, as opposed to those involving transference of heterochromatic portions of the Y long arm to D- and G-group acrocentrics, lead to reduced fertility ranging from oligo- to azoospermia.[125] Pairing disturbances in the form of univalents and unequal bivalents have been seen at metaphase I with the light microscope, but little information from surface spreading is as yet available.

A mechanism to explain the reduced fertility in the X-autosome male has been proposed by Lifschytz and Lindsley,[101] who found that 80% of such translocation carriers in *Drosophila* were sterile. To enable normal spermatogenesis to proceed, it was deemed essential that the XY bivalent progress asynchronously through prophase in relation to the autosomes. Any disturbance in this system, such as the bringing of the X chromosome into contact with an autosome via a translocation, would lead to maturation arrest. Extended to mouse and man, if the normally inactive X chromosome were to become transcriptionally active at an inappropriate time, disturbances in germ cell maturation would lead to eventual infertility.

B. Autosome-Autosome Translocations
1. Reciprocal

Translocations involving only autosomes also cause sterility in man. In the mouse, it was observed that an excess number of translocations with chain configurations at metaphase I of meiosis was associated with infertility.[126] Further analysis has shown that the location of one breakpoint near a telomere was the causal factor. This produces an asymmetrical quadrivalent with at least one short arm. Failure of synapsis within this short arm at pachytene will then lead to a chain configuration at metaphase I. From 11 human reciprocal translocations reviewed by Chandley,[42] one azoospermic male had 100% of chains, while three males with sperm counts of 20×10^6/ml or more had no chains and 100% ring configurations. Apart from the purely mechanical and time scheduling constraints imposed upon such a chain quadrivalent at the first meiotic spindle, Forejt[89] has proposed that quadrivalent contact with the X chromosome might be of importance regarding fertility (Figure 8c). Such contacts

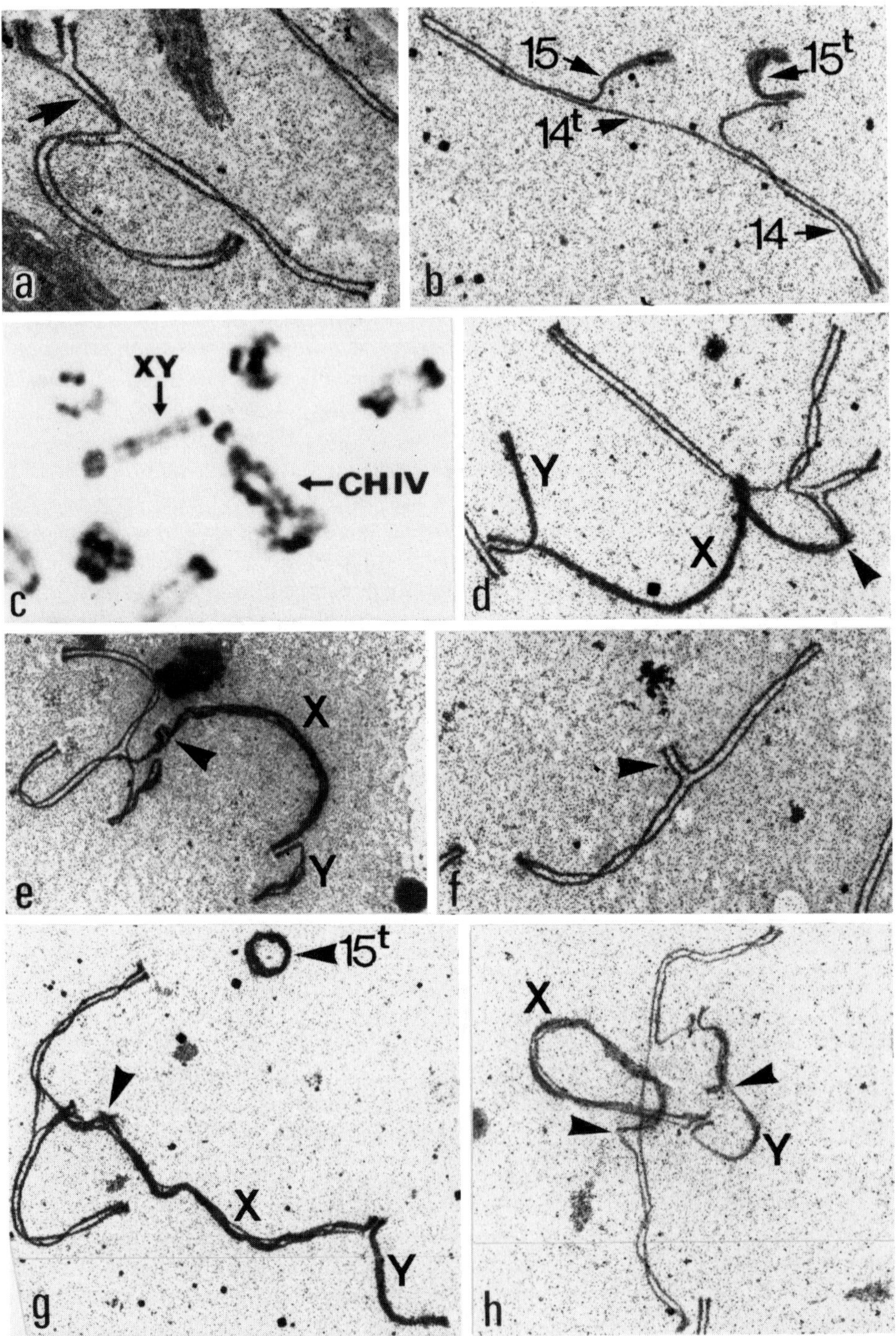

FIGURE 8.   Synaptic behavior in the T(14;15) 6 Ca male sterile mouse. (a) Fully paired quadrivalent with heterologous synapsis around breakpoints (arrow). (b) Failure of short arm synapsis leading to thickened appearance of small marker chromosome 15ᵗ and part of chromosome 15. (c) Quadrivalent/XY bivalent at MI (light microscope). (d) X contact (arrowhead) with fully paired quadrivalent. (e) X contact with thickened region of chromosome 15 (arrowhead). (f) Heterologous synapsis (arrowhead) within trivalent formed between chromosomes 14, 14ᵗ, and 15. (g) X contact with unpaired region of trivalent (arrow). Chromosome 15ᵗ showing telomere association (arrowhead). (h) X contact with chromosome 15, Y contact with 15ᵗ (arrowheads). (Figure 8c from Forejt, J., *Genetic Control of Gamete Production and Function*, Crosignani, P. G., Rubin, B. L., and Fraccaro, M., Eds., Academic Press, London, 1982, 135. With permission.)

in male sterile mice varied in number according to the type of translocation present and strain background. As an extension of the Lifschytz and Lindsley[101] hypothesis, it has been suggested that the autosomal contact brings about a reactivation of the X chromosome,[89] this being supported by the apparent reactivation of X chromosome transcription in male sterile mice.[127] The initial observation by Forejt[128] in the T(14;15)6 Ca mouse translocation, however, indicated that it was the autosomal region in contact with the X chromosome that became heteropycnotic (transcriptionally inactive) and dark staining, resembling the X chromosome itself. Such an inactivation might also be the case in an unbalanced Y-autosome translocation.[129] At pachytene, the SC formed by the Y/14 chromosome showed an extension of its Y's thickened morphology into the region of chromosome 14. Conversely, in sterile double Robertsonian-carrying male mice, autosome/sex bivalent contact results in decondensation of the X and Y chromosomes.[130] Human translocations associated with infertility are usually of the former, less complex reciprocal type. The origins of the quadrivalent/sex bivalent associations are seen at pachytene,[131] when the unpaired arms of quadrivalents which normally make contact are of a similar thickened, dark-staining morphology as the XY bivalent. However, does the thickened morphology of unpaired autosomal or sex chromosome lateral elements at pachytene bear any relationship at all to the heteropycnotic appearance of the chromatin observed at metaphase I? Thickened lateral elements generally indicate asynapsis, as in the male, where partial asynapsis of the XY bivalent contained within the sex vesicle is the norm. Wahrman and Richler, quoted in Roseman et al.,[132] suggest that although the sex chromosomes are transcriptionally inactive and appear thickened and dark staining at pachytene, they maintain their active chromatin conformation. Contact between thickened asynaptic autosomal regions and the normally asynaptic sex bivalent at prophase cannot, then, argue either for inactivation of autosomal chromatin or reactivation of X chromosome chromatin, but merely reflects a mechanism to segregate asynaptic regions occurring during pachytene in males. A preliminary investigation of T(6) Ca male mice using surface spreading at prophase, has shown a complex pattern of events to be occurring, several of which may lead to cell death. The quadrivalent is only occasionally fully paired and often shows heterologous synapsis in the region of the breakpoints (Figure 8a). Failure of synapsis and a thickened morphology in the arm concerned can occur prior to association with the XY bivalent (Figure 8b). The quadrivalent may also associate with the XY bivalent with no apparent asynapsis or thickening of autosomal material (Figure 8d). Preferential XY contact with the 15^t chromosome was observed by Forejt,[128] but contact with chromosome 15, resembling an SC, was most frequently seen in our material (Figure 8e). The small marker 15^t chromosome may fail to pair, being included in the sex vesicle as in tertiary trisomic male mice. The remaining three translocation chromosomes then pair as a trivalentlike structure, with heterologous synapsis occurring between chromosomes 14 and 15 (Figure 8f). Such trivalent-like structures may associate with the XY bivalent (Figure 8g), as do the partially asynaptic quadrivalents. Occasionally, more complex associations between quadrivalents and both X and Y chromosomes occur (Figure 8h). Obviously, a more complex synaptic pattern of autosome/XY contact can exist at prophase, compared with that seen at metaphase I. A further mechanism to explain male sterility in relation to synaptic failure was proposed by Miklos.[29] Any univalents or asynaptic regions present during meiosis would lead to a failure to saturate homologous pairing sites and to an elimination of such germ cells. Asynaptic regions are common around the breakpoints of human translocation quadrivalents and, even in contact with the sex bivalent, the initial event is an asynaptic one which may, in itself, be the primary cause of cell death. In T(6) Ca, the small marker chromosome may behave as an univalent, being included in the sex vesicle. Considering the above mechanisms of spermatogenic breakdown, we can divide human balanced reciprocal translocations into two groups.

Translocations involving acrocentrics — Human translocations involving acrocentric chromosomes, compared with those that do not, lead to increased germ cell death. Gabriel-Robez et al.[133] have shown a ratio of 1:4 for translocations with an acrocentric present, compared with nonacrocentric translocations in fertile males. This ratio changes to 1:1 in patients ascertained through subfertility. In what way might acrocentric chromosomes bring about this change? In normal males, it has been observed that the heterochromatic short arm regions of the D- and G-group acrocentrics which also carry the nucleolar organizer region associate with the sex vesicle.[134,135] It has already been shown that asynaptic regions in translocation quadrivalents associate with the XY bivalent. The further presence of acrocentric elements with an added affinity for the sex chromosomes may enhance cell disruption. Table 4 presents the available data on human translocations involving acrocentrics, and those which do not, and their levels of association with the XY bivalents seen at pachytene. Although not all authors present accurate sperm counts or details of testicular histology for translocations involving acrocentrics, increased quadrivalent/XY association is suggestive of increased levels of fertility disturbance, as reflected in the lowered sperm count.

Translocations not involving acrocentrics — The second class of human balanced reciprocal translocations shows a lower number of quadrivalent/XY associations and a generally higher sperm count (Table 4). There is some overlap between the groups, the patients with the highest XY contact among the nonacrocentric translocations having the lowest sperm counts, which fall within the acrocentric translocation range. General pairing disturbances are seen, as in the 46,XY,t(9;20) studied at Edinburgh. Asynapsis around the breakpoints (Figure 9a) is a common feature, and in the few quadrivalents that do fully pair, heterologous synapsis occurs in the central region comparable with that seen in the mouse T(6) Ca translocation (Figure 9b). Again, the main causal agent promoting contact between the quadrivalent and XY bivalent is a failure of synapsis in one arm of the former, this occurring in about 20% of the cells. The morphology of the unpaired autosomal lateral elements is again similar to the X and Y chromosomes that they contact (Figure 9c, d). There also appeared to be large numbers of germ cells degenerating before they entered the prophase stage of meiosis in this patient, suggesting that translocation-associated spermatogenic breakdown may even be initiated premeiotically.

Human female carriers of reciprocal translocations have previously been thought of as fertile. However, as discussed earlier, the female can be effectively fertile with far fewer germ cells than is the case in the male. In the mouse, Mittwoch et al.[144] have shown that for the T42H translocation causing male sterility, female germ development is also impaired, numbers of oocytes and ovary size one week after birth being much reduced. Little evidence exists for humans, but in the aneuploid XO human fetal ovaries, a premeiotic block exists with few oocytes progressing to maturity.

2. Robertsonian Translocations

Human Robertsonian translocations, most commonly involving chromosomes 13 and 14, constitute the largest group of translocation heterozygotes presenting at infertility clinics.[41] Other combinations of D- and G-group acrocentrics do occur, but with a much lower frequency. The effects of such centric fusions on fertility vary from azoospermia to normal (Table 5). Individuals within families carrying the same translocation may vary from fertile to sterile, but as Rosenmann et al.[132] have pointed out, such carriers will inherit different combinations of the normal chromosomes 13 and 14, which combine with the centric fusion chromosomes to form the trivalent at meiosis. Genetic background, then, may also be of importance in such rearrangements. As in the reciprocal translocations, contact with the XY sex bivalent is a feature at pachytene. It can vary from 0%[145] to 75%[132] there being a suggestion of a trend toward reduced fertility with increasing contact (Table 5). There are exceptions, however, the case of Johannison et al.[138] having 20% trivalent contact with the XY bivalent, but the highest recorded sperm count. In a 45,XY,t(14;22) translocation studied

Table 4

RELATIONSHIP BETWEEN BALANCED RECIPROCAL TRANSLOCATION TYPE, XY ASSOCIATIONS, AND FERTILITY

Karyotype	Method of analysis	Association of quadrivalent with XY (%)	Testicular histology	Seminal analysis	Ref.
Involving acrocentrics					
46,XY,t(9;12;13)(q22;q22;q23)	EM	0.0	Essentially normal	No count	136
46,XY,t(9;21)(q33.2;q22.1)	EM	11.0	Arrest at early spermatocyte stage	Oligospermia 27.9 $\times$ 10^6/ml	137
46,XY,t(9;15)(p22;q15)		42.5	Some arrest at spermatocyte stage	Oligospermia 32.6 $\times$ 10^6/ml	138
46,XY,t(14;21)(q13;p13)		65.5	Complete arrest of spermatogenesis	Azoospermic	138
46,XY,t(17;21)(p13pq11)	EM	70	Not recorded	Both azoospermic	
46,XY,t(19;21)(p13;q11.1)	EM	(Both patients combined)			133
46,XY,t(17;21)(p13;q11)	LM[a]	77.7	Arrest at spermatid stage	Oligospermia 3.9 $\times$ 10^6/ml	139
Not involving acrocentrics					
46,XY,t(4;17)(q21.1;q24)	EM	0.0	Not recorded	45 $\times$ 10^6/ml	140
46,XY,t(2;4;9)(p13;q25;p12)	LM[(s)] + EM	1.2	Not recorded	Normospermia 122 $\times$ 10^6/ml	141
46,XY,t(4;16)(p14;p11.2)	EM	1.6	Some arrest at early spermatocyte stage	Normospermia	
46,XY,t(7;20)(q33;p13)	EM	5.2	Reduced spermatogenesis	No count	137
46,XY,t(9;20)(q34;q11)	EM	20.0	Reduced spermatogenesis Few spermatozoa	Oligospermia 9.2 $\times$ 10^6/ml	142
46,XY,t(10;17)	EM	31.0	Maturation arrest	No count	143

Note: EM, Electron microscope; LM[a], Light microscope, air-dried preparations; LM[s], Light microscope, surface-spread preparations.

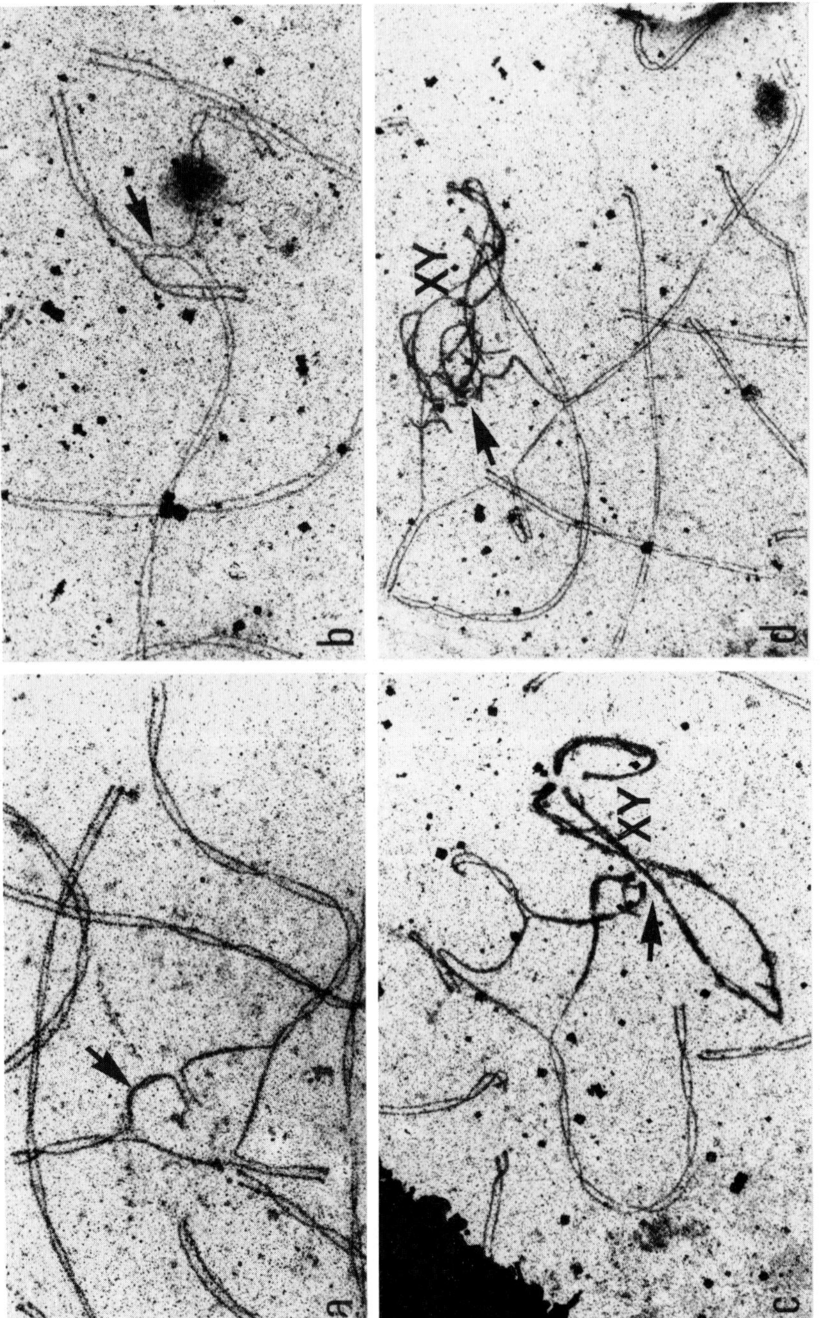

FIGURE 9. Quadrivalent morphology in a human 46,XY,t(9;20) reciprocal translocation. (a) Quadrivalent showing incomplete synapsis in the vicinity of breakpoints (arrow). (b) Heterologous synapsis of the quadrivalent lateral elements around the breakpoints (arrow). (c,d) Quadrivalent/XY bivalent association (arrow). Thickened, unpaired arms of the quadrivalent contact an unpaired X and Y in (c) and a Type VI sex bivalent in (d). (From Chandley, A. C., Speed, R. M., McBeath, S., and Hargreave, T. B., *Cytogenet. Cell Genet.*, 41, 145, 1986. With permission.)

Table 5
RELATIONSHIP BETWEEN ROBERTSONIAN TRANSLOCATIONS AND XY ASSOCIATION WITH REGARD TO FERTILITY IN THE HUMAN MALE

Karyotype	Method of analysis	Association of trivalent with XY (%)	Testicular histology	Seminal analysis	Ref.
45,XY,t(13q;14q)		No association in either case	Not recorded	Oligospermia 27.6 $\times$ 10^6/ml	
45,XY,t(13q;14q)	LM^s + EM			Oligospermia 22.4 $\times$ 10^6/ml	145
45,XY,t(13q;14q)	EM	5.3	Reduced spermatogenesis	Oligospermia 14.0 $\times$ 10^6/ml	137
45,XY,t(13q;14q)	EM	8.8	Arrest of spermatogenesis	Azoospermia	
45,XY,t(14;22)(p11;q11.1)	EM	12.0	Not recorded	Oligospermia 15 $\times$ 10^6/ml	140
45,XY,t(13q;14q)	EM	20.0	Normal	173 $\times$ 10^6/ml	138
45,XY,t(13q;14q)	LM^a	61.0	Late arrest; only a few spermatozoa	Severe oligospermia 1.6 $\times$ 10^6/ml	146
45,XY,t(14q;21q)		64 % at LM	Arrest at spermatocyte/ spermatid stage	Severe oligospermia (no count)	
45,XY,t(14q;21q)	LM^D + EM	(Both cases combined) 75% at EM (one case only)			132

Note: EM, Electron microscope; LM^a, Light microscope, air-dried preparation; LM^s, Light microscope, surface-spread preparations.

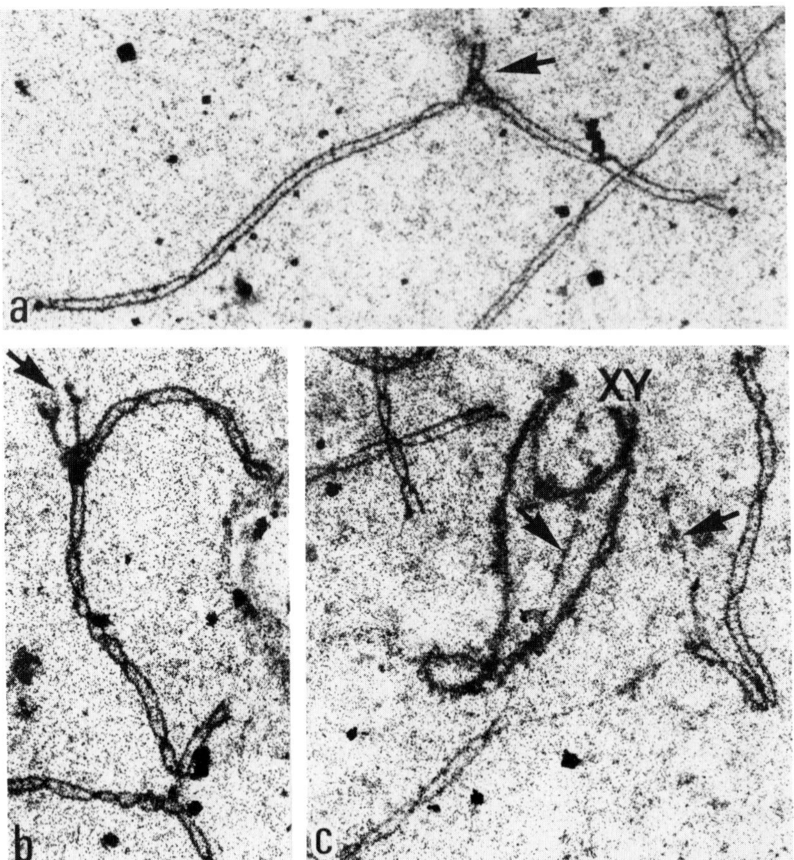

FIGURE 10.   Trivalent morphology in a human 45,XY,t(14;22) Robertsonian translocation.
(a) Trivalent showing heterologous synapsis of the normal chromosome 14 and 22 short arms
(arrow). (b) Trivalent showing lack of short-arm synapsis (arrow). (c) Trivalent associated
with the XY bivalent via the unpaired short arms (arrows).

by surface spreading at Edinburgh, 42.4% of the trivalents were fully paired, including
heterologous synapsis of the acrocentric short arms (Figure 10a). In the remaining 57.6%
of trivalents, the acrocentric short arms were asynapsed (Figure 10b). Within this category
of trivalent, 21.1% were associated with the sex bivalent via the unpaired short arms,
representing 12.0% of the pachytene nuclei overall (Figure 10c). Balanced reciprocal translocations involving acrocentrics show increased germ cell death compared with nonacrocentric translocations, and this is taken one stage further in Robertsonian fusions, where the
short arms of two differing acrocentric chromosomes are present. Again, is the problem of
XY contact with the trivalent as proposed by Forejt[89] the main cause of maturation breakdown,
or does the presence of asynaptic telomere regions create the primary defect as based on
the theory of Miklos?[29] Such simple alternatives are unlikely to resolve fully the mechanism
of translocation-induced infertility in mouse or man. Translocations appear to induce both
premeiotic breakdown, and postmeiotic segregational losses; at the pachytene level, further
biochemical advances are required for a clear understanding of the underlying problems.

VI. INVERSIONS, SYNAPTIC ADJUSTMENT, AND FERTILITY

Pericentric inversions in humans, as in other species, would be expected to reduce fertility
by the production of unbalanced gametes (duplication-deficiency) due to crossing over within

the inverted segment. In general, such chromosomal rearrangements have been ascertained through the identification of children born with congenital malformations. However, studies of synaptic behavior at prophase and metaphase I of meiosis in the male have relied on material obtained from subfertility clinics. Within the limited number of patients reported in this manner (Table 6), human chromosome 1 inversions appear most frequently. An interesting observation from surface-spread spermatocytes is the apparent rarity of classical, fully synapsed loop formation in both human and animal species.[150] Instead, the inverted region may show partial loop formation or extensive asynapsis, with the typical thickened and dark staining appearance of pairing failure (Figure 11a,b). Finally, the inverted region may synapse heterologously throughout its length, forming an SC indistinguishable from normal.

In mice heterozygous for tandem duplications[97] or paracentric inversions,[21] a phenomenon of synaptic adjustment has been described. Fully synapsed inversion loops decrease in size from early pachytene, until by mid/late pachytene the loops have resolved themselves by secondary heterologous pairing and resemble normal SCs. In some human inversion cases,[148,149] it has been reported that heterologous pairing along the inversion occurs at the late prophase stage when synaptic adjustment normally occurs. However, in the Inv(1) studied by Chandley et al.,[150] fully heterosynapsed inversion bivalents were seen in the earliest type O spermatocytes. This was also the case in a patient with an Inv(13) reported by Saadallah and Hultén.[147] Such an early zygotene initiation of heterologous synapsis has been observed in *Peromyscus sitkensis* (deer mouse),[152] which may also be a general characteristic of this species.

In what way, then, do inversions influence fertility in human males? The asynapsis seen in the inversion carrying bivalent No. 1 of man may primarily arise from the presence of a large heterochromatic block delaying synapsis.[150,151] However, apart from the Inv(13) case,[147] little evidence of contact between the asynaptic region and the XY bivalent was observed (Table 6). As previously discussed, contact with the sex bivalent usually involves unpaired telomeric regions of chromosomes present in univalents or the asynaptic arms of translocation quadrivalents, and infertility, as proposed by Forejt,[128] is unlikely in inversion carrying patients. A further general asynapsis of both autosomes and sex chromosomes was seen in two inversion carriers.[150,151] Such an interchromosomal effect might lead to cell death, according to the theory of Miklos.[29]

Heterologous synapsis may, on the other hand, rescue a proportion of the developing spermatocytes in a similar way to that reported for the self-synapsis of the single X chromosome in female XO mice.[20] In the Inv(13) case reported by Saadallah and Hultén,[147] 74.2% of spermatocytes showed heterologous pairing of the inversion bivalent, and while no sperm count was available, testicular histology was described as essentially normal. For the remaining cases of Inv(1), heterologous synapsis was reduced to levels ranging from zero to 27.4%. Here, fertility was reduced to severe oligo- or azoospermia. Further case studies will be necessary to determine if this relationship is a valid one. A final observation of synaptic behavior in human inversion carriers suggests that early heterologous pairing will prevent loop formation in which crossing over could occur. In one human case,[150] this is reflected in the reduced chiasma frequency seen at meiosis I both within the inversion bivalent and especially within the inversion region. It is possible that far fewer recombinant genotypes resulting from unbalanced gametes of the duplication/deletion varieties will be seen among the offspring of human inversion carriers.

VII. GENETIC CONTROL OF ASYNAPSIS AND DESYNAPSIS

As all cellular processes will be ultimately controlled by the genetic constitution of the individual, the normal development of the germ cells will be no exception to the rule. The

Table 6

RELATIONSHIP BETWEEN HETEROLOGOUS PAIRING, XY CONTACT, AND FERTILITY IN HUMAN MALE INVERSION CARRIERS

Karyotype	Method of analysis	Heterologous pairing of inversion region		XY-inversion association	Autosomal asynapsis	Testicular histology	Seminal analysis	Ref.
		%	No. cells					
46,XY,Inv(13)(p12q14)	EM	74.2	121/163	11.0	Not recorded	Essentially normal, but signs of increased de-generation	Not available	147
46,XY,Inv(1)(p32q21), Inv(9)(p11q12)	LM[a]	22.6 (Inv 1) 27.4 (Inv 9)	26/115 25/91	No association	Not recorded	General arrest at sper-matid level, but $^1/_3$ tu-bules show prespermatid arrest	Oligospermia 6 to 14×10^6/ ml	148
46,XY,Inv(1)(p32q12)	EM	16.7	5/30	No association	Not recorded	Not recorded	Oligospermia 1.2×10^6/ml	149
46,XY,Inv(1)(p31q43)	EM	16.4	11/67	No association	45%	Arrest at spermatid level	Oligospermia 1.0×10^6/ml	150
46,XY,Inv(1)(p32q42)	EM	0.0	0/134	No association	20%	Complete arrest at late 1st spermatocyte level	Azoospermia	151

Note: EM, Electron microscope analysis of surface-spread spermatocytes; LM[a], Light-microscopic analysis of air-dried spermatocytes.

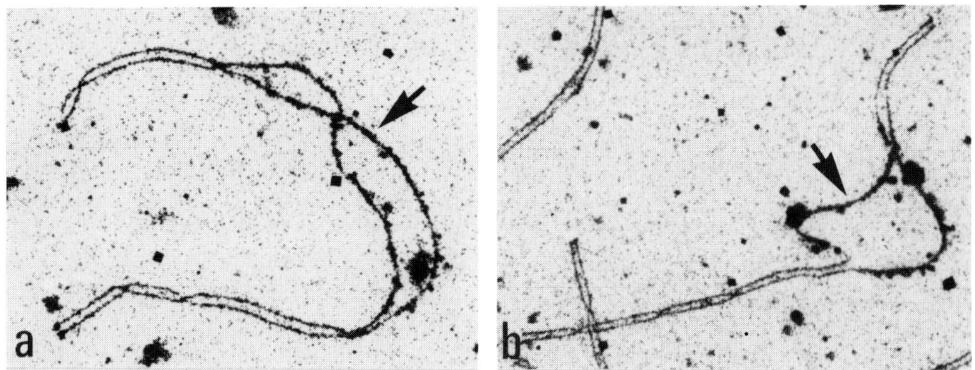

FIGURE 11. Bivalent morphology in a human 46,XY,Inv(1). (a) Inversion represented by extensive interstitial asynapsis (arrow). (b) Progressive heterologous synapsis within the inversion has reduced the region of asynapsis (arrow). (From Chandley, A. C., McBeath, S., Speed, R. M., Yorston, L., and Hargreave, T. B., *J. Med. Genet.*, 24, 325, 1987. With permission.)

estimated number of 50,000 gene loci present in the human genome[153] far outnumbers their known function either at the cellular or developmental level. Control of the highly specialized meiotic system leading to male and female germ cell production must account for at least a proportion of them.[154] The formation and control of specialized organelles, such as the SC and the associated process of recombinational synapsis, will also involve the operation of many structural and regulatory genes specifically programmed to function at precise, but differing times in males and females. Genes controlling the normally functioning meiotic system will be difficult to identify, and most information has come from the effects of mutation in species as diverse as *Saccharomyces*[155] and *Drosophila*.[156] The genes controlling meiosis in *Drosphila* are largely independent in males and females, few mutations producing effects in both sexes. In addition, most data have come from the female since crossing over is restricted to that sex in *Drosophila*. The mutant genes c3G and ord,[155] function at the prepachytene stage and lead to abnormal SC formation, reduced recombination, and increased levels of nondisjunction. In most cases, the effects of such mutants are complex and the function of the normal allele remains difficult to define at the level of gene products and function within the nucleus. Even in the mouse, the most widely studied experimental mammal, only about 25 genes are known to have a serious effect on fertility, which is less than half of those known to influence coat color.[120] Such fertility genes may affect early germ cell maturation, dominant spotting (W^v) causing a failure of the mitotic division in the primordial germ cells and their migration to the germinal ridge. The hop-sterile gene (*hop*) affects late maturation of the spermatozoa, no sperm tails being formed.[120] Overall genetic background can also be of importance, Forejt[157] describing a locus male sterility-1 (Mst-1) which controls the extent of spermatogenic impairment in male mice carrying the T6 Ca translocation. On the inbred C57BL/6 background, T6 Ca heterozygote males have a much reduced testis weight and low sperm count.

In a single male mouse, Purnell[158] reported a block in spermatogenesis, with meiosis subsequently being achiasmatic. The defect was thought to be caused by a single recessive mutant gene. In humans, while no positive examples of meiotic mutants have been described, as in other species, they must surely exist. Studies of subfertile males occasionally identify individuals with pairing disturbances at metaphase I of meiosis. Chiasma count is reduced, resulting in the presence of univalents, single chiasma bivalents, and fragments in the majority of cells.[159] Such males are oligo- or azoospermic and it has been proposed that an asynaptic or desynaptic mutation is the causal agent.[160] Perhaps the timing of meiosis, for which many genes must exist, will be a critical factor in synaptic behavior. Even premeiotically, a major change in chromosome behavior occurs, from a cycle of mitotic divisions which increase in duration to a single meiotic division whose duration varies in males and females of the same species. The control of pairing by variations in time was originally suggested by Darlington[161] in his time limit for pairing hypothesis. A few examples of such control exist. Klein[162] observed an increased duration of meiosis in two mutant forms of *Pisum sativum*. In *Triticum aestivum*, the absence of chromosome 5B again increases the duration of meiosis.[163] For the mouse, the presence of abnormal karyotypes leads to a similar phenomenon. Spermatocytes of T70H tertiary trisomics show a delay in the transition from first to second meiotic metaphase,[164] as do germ cells of Rb7Bnr/Rbl Iem double Robertsonian heterozygotes.[89] The initial events leading to such a delay appear to be pairing failure at the first meiotic prophase and univalent association with the sex bivalent.

Even though important advances in cytological and biochemical fields have occurred, the information presently available is still inadequate to explain the synaptic processes, be they normal or abnormal. Further studies on the fine structure of the meiotic process are required, including the biochemical nature of the isolated SC[165,166] and SC associated chromatin. Novel immunological approaches[15,167] will also aid our interpretation of the synaptic behavior of the SC at pachytene and of its influence on fertility in man and mammals.

REFERENCES

1. **Darlington, C. D.,** Meiosis in perspective, *Philos. Trans. R. Soc. London Ser. B,* 277, 185, 1977.
2. **Berent, J.,** Family planning in Europe and USA in the 1970s, Comparative Studies, No. 20, *World Fertility Survey,* London, 1982.
3. **Varma, T. R.,** Infertility, *Br. Med. J.,* 294, 887, 1987.
4. **Ford, C. E. and Hamerton, J. L.,** The chromosomes of man, *Nature,* 178, 1020, 1956.
5. **Evans, E. P., Breckon, G., and Ford, C. E.,** An air-drying method for meiotic preparation from mammalian testes, *Cytogenetics,* 3, 289, 1964.
6. **Counce, S. J. and Meyer, G. F.,** Differentiation of the synaptonemal complex and the kinetochore in *Locusta* spermatocytes studied by whole-mount electron microscopy, *Chromosoma,* 44, 231, 1973.
7. **Fletcher, J. M.,** Light microscope analysis of meiotic prophase chromosomes by silver staining, *Chromosoma,* 72, 241, 1979.
8. **Chandley, A. C., Goetz, P., Hargreave, T. B., Joseph, A. M., and Speed, R. M.,** On the nature and extent of XY pairing at meiotic prophase in man, *Cytogenet. Cell Genet.,* 38, 241, 1984.
9. **Holm, P. B. and Rasmussen, S. W.,** Human meiosis. I. The human pachytene karyotype analysed by three-dimensional reconstruction of the synaptonemal complex, *Carlsberg Res. Commun.,* 42, 283, 1977.
10. **Rasmussen, S. W. and Holm, P. B.,** Human Meiosis. II. Chromosome pairing and recombination nodules in human spermatocytes, *Carlsberg Res. Commun.,* 43, 275, 1978.
11. **Bojko, M.,** Human meiosis. VIII. Chromosome pairing and formation of the synaptonemal complex in oocytes, *Carlsberg Res. Commun.,* 48, 457, 1983.
12. **Westergaard, M. and von Wettstein, D.,** The synaptinemal complex, *Ann. Rev. Genet.,* 6, 71, 1972.
13. **Gillies, C. B.,** Synaptonemal complex and chromosome structure, *Ann. Rev. Genet.,* 9, 91, 1975.
14. **Rasmussen, S. W.,** Ultrastructural studies of spermatogenesis in *Drosophila melanogaster* Meigen, *Z. Zellforsch. Mikrosk. Anat.,* 140, 125, 1973.
15. **Dresser, M. A.,** The synaptonemal complex and meiosis: an immunocytochemical approach, in *Meiosis,* Moens, P. B., Ed., Academic Press, London, 1987, chap. 8.
16. **Ratomponirina, C., Andrianivo, J., and Rumpler, Y.,** Spermatogenesis in several intra- and interspecific hybrids of the lemur *(Lemur), J. Reprod. Fertil.,* 66, 717, 1982.
17. **Moens, P. B.,** Quantitative electron microscopy of chromosome organisation at meiotic prophase, *Cold Spring Harbor Symp. Quant. Biol.,* 38, 99, 1973.
18. **Gillies, C. B.,** Ultrastructural studies of the association of homologous and non-homologous parts of chromosomes in the mid-prophase of meiosis in *Zea mays, Maydica,* 28, 265, 1983.
19. **Von Wettstein, D., Rasmussen, S. W., and Holm. P. B.,** The synaptonemal complex in genetic segregation, *Ann. Rev. Genet.,* 18, 331, 1984.
20. **Speed, R. M.,** Oocyte development in XO foetuses of man and mouse: the possible role of heterologous X-chromosome pairing in germ cell survival, *Chromosoma,* 94, 115, 1986.

21. **Moses, M. J., Poorman, P. A., Roderick, T. H., and Davisson, M. T.,** Synaptonemal complex analysis of mouse chromosomal rearrangements. IV. Synapsis and synaptic adjustment in two paracentric inversions, *Chromosoma,* 84, 457, 1982.
22. **Moses, M. J., Poorman, P. A., Russell, L. B., Cacheiro, N. L. Roderick, T. H., and Davisson, M. T.,** Synaptic adjustment: two pairing phases in meiosis?, *J. Cell Biol.,* 79, 123a, 1978.
23. **Sadasivaiah, R. S. and Kasha, K. J.,** Meiosis in haploid barley — an interpretation of non-homologous chromosome associations, *Chromosoma,* 35, 247, 1971.
24. **Gillies, C. B.,** Ultrastructural analysis of maize pachytene karyotypes by three-dimensional reconstruction of the synaptonemal complexes, *Chromosoma,* 43, 145, 1973.
25. **Westergaard, M. and von Wettstein, D.,** Studies on the mechanism of crossing-over. IV. The molecular organisation of the synaptinemal complex in *Neottiella* (Cooke) Saccardo (Ascomycetes), *C.R. Trav. Lab. Carlsberg,* 37, 239, 1970.
26. **Moses, M. J. and Poorman, P. A.,** Synapsis, synaptic adjustment and DNA synthesis in mouse oocytes, *Chromosomes Today,* 8, 90, 1984.
27. **Riley, R. and Flavell, R. B.,** A first view of the meiotic process, *Philos. Trans. R. Soc. London Ser. B,* 277, 191, 1977.
28. **Chandley, A. C.,** A model for effective pairing and recombination at meiosis based on early replicating sites (R-bands) along chromosomes, *Hum. Genet.,* 72, 50, 1986.
29. **Miklos, G. L. G.,** Sex chromosome pairing and male fertility, *Cytogenet. Cell Genet.,* 13, 558, 1974.
30. **Burgoyne, P. S. and Baker, T. G.,** Meiotic pairing and gametogenic failure, in *Controlling Events in Meiosis,* Evans, C. W. and Dickinson, H. G., Eds., Company of Biologists, Cambridge, 1984, 349.
31. **Burgoyne, P. S. and Biddle, F. G.,** Spermatocyte loss in XYY mice, *Cytogenet. Cell Genet.,* 28, 143, 1980.
32. **McIlree, M. E., Selby-Tulloch, W., and Newsam, J. E.,** Studies on human meiotic chromosomes from testicular tissue, *Lancet,* 1, 679, 1966.
33. **Kjessler, B.,** Chromosomal constitution and male reproductive failure, in *Male Fertility and Sterility,* Mancini, R. E. and Martini, L., Eds., Academic Press, New York, 1974, 231.
34. **Baker, T. G.,** A quantitative and cytological study of germ cells in human ovaries, *Proc. R. Soc. London Ser. B,* 158, 417, 1963.
35. **Speed, R. M.,** The prophase stages in human foetal oocytes studied by light and electron microscopy, *Hum. Genet.,* 69, 69, 1985.
36. **Baker, T. G.,** Oogenesis and ovulation, in *Reproduction in Mammals,* Vol. 1, Austin, C. R. and Short, R. V., Eds., Cambridge University Press, London, 1972, chap. 2.
37. **Monesi, V.,** Spermatogenesis and the spermatozoa, in *Reproduction in Mammals, Vol. 1,* Austin, C. R. and Short, R. V., Eds., Cambridge University Press, London, 1972, chap. 3.
38. **Skakkebaek, N. E., Bryant, J. I., and Philip, J.,** Studies on meiotic chromosomes in infertile men and controls with normal karyotype, *J. Reprod. Fertil.,* 35, 23, 1973.
39. **Chandley, A. C.,** The chromosomal basis of human infertility, *Br. Med. Bull.,* 35, 181, 1979.
40. **Koulischer, L. and Schoysman, R.,** Etude des chromosomes mitotiques et méiotiques chez les hommes infertiles, *J. Génét. Hum.,* 23, 50, 1975.
41. **Tiepolo, L., Zuffardi, O., Fraccaro, M., and Giarola, A.,** Chromosome abnormalities and male infertility, in *Oligozoospermia: Recent Progress in Andrology,* Frajese, G., Hafez, E. S. E., Conti, C., and Fabbrini, A., Eds., Raven Press, New York, 1981, 233.
42. **Chandley, A. C.,** Infertility and chromosomal abnormality, in *Oxford Reviews of Reproductive Biology,* Vol. 6, Clarke, J. R., Ed., Oxford University Press, Oxford, 1984, chap. 1.
43. **Chandley, A. C.,** Human meiotic studies, in *Modern Trends in Human Genetics,* 2nd ed., Emery, A. E. H., Ed., Butterworths, London, 1975, chap. 2.
44. **Chandley, A. C. and Fletcher, J.,** Centromere staining at meiosis in man, *Humangenetik,* 18, 247, 1973.
45. **Hultén, M.,** Chiasma distribution in the normal human male, *Hereditas,* 76, 55, 1974.
46. **Solari, A. J. and Tres, L.,** The three-dimensional reconstruction of the XY chromosomal pair in human spermatocytes, *J. Cell Biol.,* 45, 43, 1970.
47. **Tres, L. L.,** Nucleolar RNA synthesis of meiotic prophase spermatocytes in the human testis, *Chromosoma,* 53, 141, 1975.
48. **Hungerford, D. A.,** Chromosome structure and function in man. I. Pachytene mapping in the male, improved methods and general discussion of initial results, *Cytogenetics,* 10, 23, 1971.
49. **Ambros, P. F. and Sumner, A. T.,** Correlation of pachytene chromomeres and metaphase bands of human chromosomes, and distinctive properties of telomere regions, *Cytogenet. Cell Genet.,* 44, 223, 1987.
50. **Ohno, S., Makino, S., Kaplan, W. D., and Kinosita, R.,** Female germ cells of man, *Exp. Cell Res.,* 24, 106, 1961.
51. **Kindred, J. F.,** The chromosomes of the ovary of the human fetus, *Anat. Res.,* 147, 295, 1963.
52. **Manotaya, T. and Potter, E. L.,** Oocytes in prophase in meiosis from squash preparations of human fetal ovaries, *Fertil. Steril.,* 14, 378, 1963.

53. **Luciani, J. M. and Stahl, A.,** Etude des states de début de la meiose chez l'ovocyte foetal humain, *Bull. Assoc. Anat.,* (Nancy), 15, 445, 1971.

54. **Henderson, S. A. and Edwards, R. G.,** Chiasma frequency and maternal age in mammals, *Nature,* 218, 22, 1968.

55. **Jagiello, G. M., Ducayen, M. B., Miller, W. A., Lin, J. S., and Fang, J. S.,** A cytogenetic analysis of oocytes from *Macaca mulatta* and *Nemestrina* matured *in vitro, Humangenetik,* 18, 117, 1973.

56. **Edwards, R. G.,** Maturation *in vitro* of human ovarian oocytes, *Lancet,* 2, 926, 1965.

57. **Jagiello, G., Karnicki, J., and Ryan, R. J.,** Superovulation with pituitary gonadotrophins. Method for obtaining metaphase figures in human ova, *Lancet,* 1, 178, 1968.

58. **Yuncken, C.,** Meiosis in the human female, *Cytogenetics,* 7, 234, 1968.

59. **Chandley, A. C.,** Culture of mammalian oocytes, *J. Reprod. Fertil. Suppl.,* 14, 1, 1971.

60. **Uebele-Kallhardt, B. and Knörr, K.,** Meiotische Chromosomen der Frau, *Humangenetik,* 12, 182, 1971.

61. **Polani, P., Dewhurst, J. Sir, Fergusson, I., and Kelberman, J.,** Meiotic chromosomes in a female with primary trisomic Down's syndrome, *Hum. Genet.,* 62, 277, 1982.

62. **Polani, P. E.,** Centromere localization at meiosis and the position of chiasma in the male and female mouse, *Chromosoma,* 36, 343, 1972.

63. **Jacobs, P. A.,** unpublished data.

64. **Moses, M. J., Counce, S. J., and Paulson, D. F.,** Synaptonemal complex complements of man in spreads of spermatocytes, with details of the sex chromosome pair, *Science,* 187, 363, 1975.

65. **Solari, A. J.,** Synaptonemal complexes and associated structures in microspread human spermatocytes, *Chromosoma,* 81, 315, 1980.

66. **Sumner, A. T. and Speed, R. M.,** Immunocytochemical labelling of the kinetochore of human synaptonemal complexes, and the extent of pairing of the X and Y chromosomes, *Chromosoma,* 95, 359, 1987.

67. **Koller, P. C. and Darlington, C. D.,** The genetical and mechanical properties of the sex chromosomes. I. *Rattus norvegicus,* ♂, *J. Genet.,* 29, 159, 1934.

68. **Ohno, S., Kaplan, W., and Kinosita, R.,** Do XY- and O-sperm occur in *Mus musculus?, Exp. Cell Res.,* 18, 382, 1959.

69. **Burgoyne, P. S.,** Genetic homology and crossing-over in the X and Y chromosomes of mammals, *Hum. Genet.,* 61, 85, 1982.

70. **Cattanach, B. M., Pollard, C. E., and Hawkes, S. G.,** Sex-reversed mice: XX and XO males, *Cytogenetics,* 10, 318, 1971.

71. **Singh, L. and Jones, K. W.,** Sex reversal in the mouse *(Mus musculus)* is caused by recurrent non reciprocal cross over involving the X and an aberrant Y chromosome, *Cell,* 28, 205, 1982.

72. **Evans, E. P., Burtenshaw, M. D., and Cattanach, B. M.,** Meiotic crossing-over between the X and Y chromosomes of male mice carrying the sex reversing (Sxr) factor, *Nature,* 300, 443, 1982.

73. **Chandley, A. C. and Fletcher, J. M.,** Meiosis in Sxr male mice. I. Does a Y-autosome rearrangement exist in sex-reversed (Sxr) mice?, *Chromosoma,* 81, 9, 1980.

74. **Evans, E. P., Burtenshaw, M. D., and Brown, B. B.,** Meiosis in Sxr male mice. II. Further absence of cytological evidence for a Y-autosome rearrangement in sex-reversed (Sxr) mice, *Chromosoma,* 81, 19, 1980.

75. **Chandley, A. C. and Speed, R. M.,** Cytological evidence that the Sxr fragment of XY, Sxr mice pairs homologously at meiotic prophase with the proximal testis-determining region, *Chromosoma,* 95, 345, 1987.

76. **Beechey, C. V.,** X-Y chromosome dissociation and sterility in mouse, *Cytogenet. Cell Genet.,* 12, 60, 1973.

77. **Chandley, A. C., Maclean, N., Edmond, P., Fletcher, J., and Watson, G. S.,** Cytogenetics and infertility in man. II. Testicular histology and meiosis, *Ann. Hum. Genet.,* 40, 165, 1976.

78. **Ashley, T.,** A re-examination of the case for homology between the X and Y chromosomes of mouse and man, *Hum. Genet.,* 67, 372, 1984.

79. **Miller, O. J., Drayna, D., and Goodfellow, P.,** Report of the committee on the genetic constitution of the X and Y chromosomes, Human Gene Mapping 7: Seventh International Workshop on Human Gene Mapping, *Cytogenet. Cell Genet.,* 37, 176, 1984.

80. **Rasmussen, S. W. and Holm, P. B.,** The synaptonemal complex, recombination nodules and chiasmata in human spermatocytes, in *Controlling Events in Meiosis,* Evans, C. W. and Dickenson, H. G., Eds., Company of Biologists, Cambridge, 1984, 271.

81. **Ashley, T.,** Meiotic behaviour of sex chromosomes: what is normal?, in *Chromosomes Today,* Vol. 9., Stahl, A., Luciani, J. M., and Vagner-Capodano, A. M., Eds., Allen and Unwin, London, 1987, 184.

82. **Goodfellow, P., Banting, G., Sheer, D., Ropers, H. H., Caine, A., Ferguson-Smith, M. A., Povey, S., and Voss, R.,** Genetic evidence that a Y-linked gene in man is homologous to a gene on the X chromosome, *Nature,* 302, 346, 1983.

83. **Cooke, H-J., Brown, W. R. A., and Rappold, G. A.,** Hypervariable telomeric sequences from the human sex chromosomes are pseudoautosomal, *Nature,* 317, 687, 1985.

84. **Simmler, M-C., Rouyer, F., Vergnand, G., Nystrom-Lahti, M., Ngo, K. Y., de la Chapelle, A., and Weissenbach, J.,** Pseudoautosomal DNA sequences in the pairing region of the human sex chromosomes, *Nature,* 317, 692, 1985.

85. **Goodfellow, P., Darling, S., and Wolfe, J.,** The human Y chromosome, *J. Med. Genet.,* 22, 329, 1985.

86. **Ferguson-Smith, M. A.,** X-Y chromosomal interchange in the aetiology of true hermaphroditism and of XX Klinefelter's syndrome, *Lancet,* 2, 475, 1966.

87. **Rouyer, F., Simmler, M-C., Johnsson, C., Vergnaud, G., Cooke, H. J., and Weissenbach, J.,** A gradient of sex linkage in the pseudoautosomal region of the human sex chromosome, *Nature,* 319, 291, 1986.

88. **Beaumont, H. M. and Mandl, A. M.,** A quantitative and cytological study of the oogonia and oocytes in the foetal and neonatal rat, *Proc. R. Soc. London Ser. B,* 155, 557, 1962.

89. **Forejt, J.,** X-Y involvement in male sterility caused by autosome translocations — a hypothesis, in *Genetic Control of Gamete Production and Function,* Crosignani, P. G., Rubin, B. L., and Fraccaro, M., Eds., Academic Press, London, 1982, 135.

90. **Speed, R. M.,** The possible role of meiotic pairing anomalies in the atresia of human fetal oocytes, *Hum. Genet.,* 78, 260, 1988.

91. **Speed, R. M.,** Prophase pairing in a mosaic 18p-;iso 18q human female fetus studied by surface spreading, *Hum. Genet.,* 72, 256, 1986.

92. **Kurilo, L. F.,** Oogenesis in antenatal development in man, *Hum. Genet.,* 57, 86, 1981.

93. **Mahadevaiah, S. and Mittwoch, U.,** Synaptonemal complex analysis in spermatocytes and oocytes of tertiary trisomic Ts(5^{12}) 31H mice with male sterility, *Cytogenet. Cell Genet.,* 41, 169, 1986.

94. **Speed, R. M.,** Meiotic configurations in female trisomy 21 fetuses, *Hum. Genet.,* 66, 176, 1984.

95. **Wallace, B. M. N. and Hultén, M. A.,** Triple chromosome synapsis in oocytes from a human fetus with trisomy 21, *Ann. Hum. Genet.,* 47, 271, 1983.

96. **Moses, M. J.,** Microspreading and the synaptonemal complex in cytogenetic studies, in *Chromosomes Today,* Vol. 6., De la Chapelle, A. and Sorsa, M., Eds., Elsevier/North Holland, Amsterdam, 1977, 71.

97. **Moses, M. J. and Poorman, P. A.,** Synaptonemal complex analysis of mouse chromosome rearrangements. II. Synaptic adjustment in tandem duplications, *Chromosoma,* 81, 519, 1981.

98. **Johannisson, R., Gropp, A., Winking, H., Coerdt, W., Rehder, H., and Schwinger, E.,** Down's syndrome in the male, reproductive pathology and meiotic studies, *Hum. Genet.,* 63, 132, 1983.

99. **Stearns, P. E., Droulard, K. E., and Stahhar, F. H.,** Studies bearing on fertility of male and female mongoloids, *Am. J. Ment. Defic.,* 65, 37, 1960.

100. **De Boer, P. and Branje, H. E. B.,** Association of the extra chromosome of tertiary trisomic male mice with the sex chromosome during first meiotic prophase and its significance for impairment of spermatogenesis, *Chromosoma,* 73, 369, 1979.

101. **Lifschytz, E. and Lindsley, D. L.,** The role of X-chromosome inactivation during spermatogenesis, *Proc. Natl. Acad. Sci. U.S.A.,* 69, 182, 1972.

102. **Speed, R. M.,** Abnormal RNA synthesis in sex vesicles of tertiary trisomic male mice, *Chromosoma,* 93, 267, 1986.

103. **Jagiello, G.,** Reproduction in Down syndrome, in *Trisomy 21 (Down syndrome) — Research Perspective,* De la Cruz, F. F. and Gerald, P. S., Eds., University Park Press, Baltimore, 1980, 151.

104. **Russell, P. and Altschuler, G.,** The ovarian dysgenesis of trisomy 18, *Pathology,* 7, 149, 1975.

105. **Peters, H., Byskov, A. G., and Grimsted, J.,** The development of the ovary during childhood in health and disease, in *Functional Morphology of the Human Ovary,* Coutts, J. R. T., Ed., MTP Press, Lancaster, England, 1981, 26.

106. **Coerdt, W., Rehder, H., Gausmann, I., Johannisson, R., and Gropp, A.,** Quantitative histology of human fetal testes in chromosomal disease, *Pediatr. Pathol.,* 3, 245, 1985.

107. **Skakkebaek N. E., Hultén, M., Jacobson, P., and Mikkelsen, M.,** Quantification of human seminiferous epithelium. II. Histological studies in eight 47,XYY men, *J. Reprod. Fertil.,* 32, 391, 1973.

108. **Evans, E. P., Ford, C. E., Chaganti, R. S. K., Blank, C. E., and Hunter, H.,** XY spermatocytes in an XYY male, *Lancet,* 1, 719, 1970.

109. **Chandley, A. C., Fletcher, J., and Robinson, J. A.,** Normal meiosis in two 47,XYY men, *Hum. Genet.,* 33, 231, 1976.

110. **Berthelsen, J. G., Skakkebaek, N. E., Perbøll, O., and Neilson, J.,** Electron microscopical demonstration of the X and Y chromosome in spermatocytes from human XYY males, in *Development and Function of Reproductive Organs,* Byskov, A. G. and Peters, H., Eds., Elsevier/North Holland, 1981, 328.

111. **Hultén M. and Pearson, P. L.,** Fluorescent evidence for spermatocytes with two Y chromosomes in an XYY male, *Ann. Hum. Genet.,* 34, 273, 1971.

112. **Brook, J. D.,** X-chromosome segregation, maternal age and aneuploidy in the XO mouse, *Genet. Res. Camb.,* 41, 85, 1983.

113. **Carr, D. H., Haggar, R. A., and Hart, A. G.,** Germ cells in the ovaries of XO female infants, *Am. J. Clin. Pathol.,* 49, 521, 1968.

114. **Cattanach, B. M.**, XO mice, *Genet. Res.*, 3, 487, 1962.

115. **Lyon, M. F. and Hawker, S. G.**, Reproductive life span in irradiated and un-irradiated XO mice, *Genet. Res.*, 21, 185, 1973.

116. **Burgoyne, P. S. and Baker, T. G.**, Perinatal oocyte loss in XO mice and its implications for the aetiology of gonadal dysgenesis in XO women, *J. Reprod. Fertil.*, 75, 633, 1985.

117. **Holm, P. B.**, Three-dimensional reconstruction of chromosome pairing during the zygotene stage of meiosis in *Lilium longiflorum* (Thunb), *Carlsberg Res. Commun.*, 42, 103, 1977.

118. **Hobolth, P.**, Chromosome pairing in allohexaploid wheat var Chinese Spring. Transformation of multivalents into bivalents, a mechanism for exclusive bivalent formation, *Carlsberg Res. Commun.*, 46, 129, 1981.

119. **Hasenkamf, C. A.**, Synaptonemal complex formation in pollen mother cells of *Tradescantia*, *Chromosoma*, 90, 275, 1984.

120. **Searle, A. G.**, The genetics of sterility in the mouse, in *Genetic Control of Gamete Production and Function*, Crosignani, P. G., Rubin, B. L., and Fraccaro, M., Eds., Academic Press, London, 1982, 93.

121. **Madan, K., Hompes, P. G. A., Schoemaker, J., and Ford, C. E.**, X-autosome translocation with a breakpoint in Xq22 in a fertile woman and her 47,XXX infertile daughter, *Hum. Genet.*, 59, 290, 1981.

122. **Quack, B., Speed, R. M., Luciani, J. M., Noel, B., and Chandley, A. C.**, Meiotic analysis in two human reciprocal X-autosome translocations, *Cytogenet. Cell Genet.*, 48, 43, 1988.

123. **Ashley, T., Russell, L. B., and Cacheiro, N. L. A.**, Synaptonemal complex analysis of X-7 translocations in male mice: R2 and R6 quadrivalents, *Chromosoma*, 88, 171, 1983.

124. **Sarto, G. E., Therman, E., and Patau, K.**, X inactivation in man: a woman with t(Xq − ;12q +), *Am. J. Hum. Genet.*, 25, 262, 1973.

125. **Smith, A., Fraser, I. S., and Elliot, G.**, An infertile male with balanced Y;19 translocation. Review of Y autosoma translocations, *Ann. Genet.*, 22, 189, 1979.

126. **Lyon, M. F. and Meredith, R.**, Autosomal translocations causing male sterility and viable aneuploidy in the mouse, *Cytogenetics*, 5, 335, 1966.

127. **Hotta, Y. and Chandley, A. C.**, Activities of X-linked enzymes in spermatocytes of mice rendered sterile by chromosomal alterations, *Gamete Res.*, 6, 65, 1982.

128. **Forejt, J.**, Non-random association between a specific autosome and the X chromosome in meiosis of the male mouse: possible consequence of the homologous centromeres separation, *Cytogenet. Cell Genet.*, 13, 369, 1974.

129. **Ratomponirina, C., Couturier, J., Gabriel-Robez, O., Rumpler, Y., Dutrillaux, B., Croquette, M., Rabache, Q., and Leduc, M.**, Aberrations of the synaptonemal complexes in a male 46,XY, − 14, + der(14)t(Y;14), *Ann. Genet.*, 28, 214, 1985.

130. **Forejt, J.**, Meiotic studies of translocations causing male sterility in the mouse. II. Double heterozygotes for Robertsonian translocations, *Cytogenet. Cell Genet.*, 23, 163, 1979.

131. **Forejt, J., Gregorová, S., and Goetz, P.**, XY pair associates with the synaptonemal complex of autosomal male-sterile translocations in pachytene spermatocytes of the mouse *(Mus musculus)*, *Chromosoma*, 82, 41, 1981.

132. **Rosenmann, A., Wahrman, J., Richler, C., Voss, R., Persitz, A., and Goldman, B.**, Meiotic association between the XY chromosomes and unpaired autosomal elements as a cause of human male sterility, *Cytogenet. Cell Genet.*, 39, 19, 1985.

133. **Gabriel-Robez, O., Ratomponirina, C., Dutrillaux, B., Carre-Pigeon, F., and Rumpler, Y.**, Meiotic association between the XY chromosomes and the autosomal quadrivalent of a reciprocal translocation in two infertile men, 46,XY,t(19;22) and 46,XY,t(17;21), *Cytogenet. Cell Genet.*, 43, 154, 1986.

134. **Luciani, J. M.**, Les chromosomes meiotiques de l'homme. II. Le nucléole, les chiasmas. III. La stérilité masculine, *Ann. Genet.*, 13, 169, 1970.

135. **Stahl, A., Hartung, M., Devictor, M., and Berge-Lefranc, J. L.**, The association of the nucleolus and the short arm of acrocentric chromosomes with the XY pair in human spermatocytes: its possible role in facilitating sex-chromosome acrocentric translocations, *Hum. Genet.*, 68, 173, 1984.

136. **Johannisson, R., Löhrs, U., and Passarge, E.**, Chromosomal and histological studies in a familial three-way translocation 9q;13q;12q, in *9th Int. Chromosome Conf. Marseille, 1986*.

137. **Batanian, J. and Hultén, M. A.**, personal communication.

138. **Johannisson, R., Löhrs, U., Wolff, H. H., and Schwinger, E.**, Two different XY-quadrivalent associations and impairment of fertility in men, *Cytogenet. Cell Genet.*, 45, 222, 1987.

139. **Luciani, J. M., Guichaoua, M. R., Delafontaine, D., North, M. O., Gabriel-Robez, O., and Rumpler, Y.**, Pachytene analysis in a 17;21 reciprocal translocation carrier: role of the acrocentric chromosomes in male sterility, *Hum. Genet.*, 77, 246, 1987.

140. **Speed, R. M.**, unpublished data.

141. **Saadallah, N. and Hultén, M. A.**, A complex three breakpoint translocation involving chromosomes 2, 4 and 9 identified by meiotic investigations of a human male ascertained for subfertility, *Hum. Genet.*, 71, 312, 1985.

142. **Chandley, A. C., Speed, R. M., McBeath, S., and Hargreave, T. B.**, A human 9;20 reciprocal translocation associated with male infertility, *Cytogenet. Cell Genet.*, 41, 145, 1986.

143. **Hultén, M. A., Saadallah, N., and Batanian, J.**, Meiotic chromosome pairing in the human male: experience from surface spread synaptonemal complexes, *Chromosomes Today*, 9, 218, 1987.

144. **Mittwoch, U., Mahadevaiah, S., and Olive, M. B.**, Retardation of ovarian growth in male-sterile mice carrying an autosomal translocation, *J. Med. Genet.*, 18, 414, 1981.

145. **Templado, C., Vidal, F., Navarro, J., Marina, S., and Egozcue, J.**, Meiotic studies and synaptonemal complex analysis in two infertile males with a 13/14 balanced translocation, *Hum. Genet.*, 67, 162, 1984.

146. **Luciani, J. M., Guichaoua, M. R., Mattei, A., and Morazzani, M. R.**, Pachytene analysis of a man with a 13q;14q translocation and infertility. Behaviour of the trivalent and non-random association with the sex vesicle, *Cytogenet. Cell Genet.*, 38, 14, 1984.

147. **Saadallah, N. and Hultén, M. A.**, EM investigations of surface spread synaptonemal complexes in a human male carrier of a pericentric inversion inv(13)(pl2q14), the role of heterosynapsis for spermatocyte survival, *Ann. Hum. Genet.*, 50, 369, 1986.

148. **Guichaoua, M. R., Delafontaine, D., Tourelle, R., Taillemite, J. L., Morazzani, M. R., and Luciani, J. M.**, Loop formation and synaptic adjustment in a human male, heterozygous for two pericentric inversions, *Chromosoma*, 93, 313, 1985.

149. **Gabriel-Robez, O., Ratomponirina, C., Rumpler, Y., Le Marec, B., Luciani, M., and Guichaoua, M. R.**, Synapsis and synaptic adjustment in an infertile human male heterozygous for a pericentric inversion in chromosome 1, *Hum. Genet.*, 72, 148, 1986.

150. **Chandley, A. C., McBeath, S., Speed, R. M., Yorston, L., and Hargreave, T. B.**, Pericentric inversion in human chromosome 1 and the risk for male sterility, *J. Med. Genet.*, 24, 325, 1987.

151. **Batanian, J. and Hultén, M. A.**, Electron microscopic investigations of synaptonemal complexes in an infertile human male carrier of a pericentric inversion inv(1)(p32q42). Regular loop formation but defective synapsis including a possible interchromosomal effect, *Hum. Genet.*, 76, 81, 1987.

152. **Hale, D. W.**, Heterosynapsis and suppression of chiasmata within heterozygous pericentric inversions of the Sitka deer mouse, *Chromosoma*, 94, 425, 1986.

153. **Hastie, N. L. D.**, personal communication.

154. **Rees, H.**, Genotypic control of chromosome form and behaviour, *Bot. Rev.*, 27, 288, 1961.

155. **Esposito, M. S.**, Molecular mechanisms of recombination in *Saccharomyces cerevisiae:* testing mitotic and meiotic models by analysis of hypo-rec and hyper-rec mutations, in *Controlling Events in Meiosis,* Evans, C. W. and Dickinson, H. G., Eds., Company of Biologists, Cambridge, 1984, 123.

156. **Lindsley, D. L. and Sandler, L.**, The genetic analysis of meiosis in female *Drosophila melanogaster, Philos. Trans. R. Soc. London Ser. B*, 277, 295, 1977.

157. **Forejt, J.**, Hybrid sterility gene located in the T/t-H-2 supergene on chromosome 17, in *Current Trends in Histocompatibility*, Vol. 1, Reisfeld, R. A. and Ferrone, S., Eds., Plenum Press, New York, 1981, 103.

158. **Purnell, D. J.**, Spontaneous univalence at male meiosis in the mouse, *Cytogenet. Cell Genet.*, 12, 327, 1973.

159. **Thomson, E., Fletcher, J., Chandley, A. C., and Kŭcerová, M.**, Meiotic and radiation studies in four oligochiasmatic men, *J. Med. Genet.*, 16, 270, 1979.

160. **Pearson, P. L., Ellis, J. D., and Evans, H. J.**, A gross reduction in chiasma formation during meiotic prophase and a defective DNA repair mechanism associated with a case of human male infertility, *Cytogenetics*, 9, 460, 1970.

161. **Darlington, C. D.**, The prime variables of meiosis, *Biol. Rev.*, 15, 307, 1940.

162. **Klein, H. D.**, Timing anomalies during meiosis, *Pisum Newsl.*, 4, 14, 1972.

163. **Bennett, M. D. and Smith, J. B.**, The effect of polyploidy on meiotic duration and pollen development in cereal anthers, *Proc. R. Soc. London Ser. B*, 181, 81, 1972.

164. **Speed, R. M. and de Boer, P.**, Delayed meiotic development and correlated death of spermatocytes in male mice with chromosome abnormalities, *Cytogenet. Cell Genet.*, 35, 257, 1983.

165. **Ierardi, L. A., Moss, S. B., and Bellvé, A. R.**, Synaptonemal complexes are integral components of the isolated mouse spermatocyte nuclear matrix, *J. Cell Biol.*, 96, 1717, 1983.

166. **Shende, Li., Meistrich, M. L., Brock, W. A., Hsu, T. C., and Kuo, M. T.**, Isolation and preliminary characterization of the synaptonemal complex from rat pachytene spermatocytes, *Exp. Cell Res.*, 144, 63, 1983.

167. **Moens, P. B., Heyting, C., Dietrich, A. J. J., van Raamsdonk, W., and Chen, Q.**, Synaptonemal complex antigen location and conservation, *J. Cell Biol.*, 105, 93, 1987.

Chapter 2

CHROMOSOME PAIRING AND FERTILITY IN MICE

P. de Boer and J. H. de Jong

TABLE OF CONTENTS

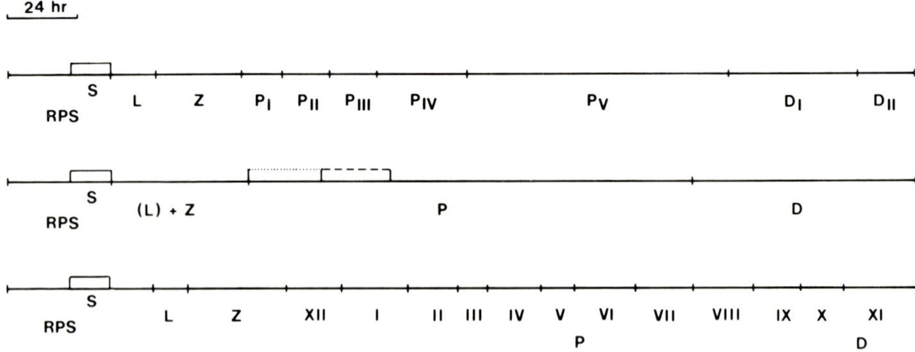

FIGURE 1.   A graphical representation of the relative and absolute durations of the nuclear stages from resting primary spermatocytes (RPS) to diplotene spermatocytes (D). The interpretations by Moses[22] (top) and Dietrich and de Boer[24] (middle) have been compared with the original data of Oakberg[18,19] (bottom). S = S-phase, L = leptotene, Z = zygotene, P = pachytene. The stages of Oakberg's scheme are in Roman numerals. The bar indicates 24 h. In the time schedule of Dietrich and de Boer,[24] the beginning of pairing for the sex chromosomes is indicated by dots; the maximal degree of pairing, by stripes.

constitution is there some suspicion that oocyte quality may be one cause for a reduction of litter size (besides the early death of OY conceptuses during the first two cleavage stages and some XO runts during the second half of gestation as the other causes). (See Reference 10 for a discussion.) A better indication of a reduced oocyte pool is a decrease of the reproductive lifespan as is seen with the XO mouse, but also with the male-sterile insertional translocation Is(7;1)40H.[15] Other candidates for this phenomenon are the female carriers of an extra chromosome (tertiary trisomics, such as Ts[1¹³]70H). Cytological confirmation of this phenomenon can be supplied by counting the number of oocytes (with few adhering cumulus cells) after pricking the ovary, assuming that antral follicles only will release their oocytes.[16] More direct evidence can be obtained from histological sections taken just before or shortly after birth and counting the number of oocytes. Appropriate controls should be included since the number of primary oocytes usually declines sharply from pachytene on[17] (see also Section III.B). As with the male carriers of chromosome mutations, we mention the indications for germ cell death when the particular mutations are discussed.

III. ZYGOTENE AND PACHYTENE IN MALE AND FEMALE MICE

A. The Male Mouse

The best way to characterize first meiotic prophase stages is to give, besides the classical designation of zygotene, pachytene, and diplotene, their time of occurrence relative to a clear point of reference, such as the division of B spermatogonia to resting primary spermatocytes (RPS), the end of premeiotic S-phase (of RPS), or the position in relation to Oakberg's scheme of spermatogenesis[18] (see Figure 1). For instance, pachytene spermatocytes of stage VII of Oakberg's scheme are an average of 220 h beyond the B spermatogonium-RPS mitosis (the duration of the stages of the Oakberg classification are given in Reference 19). With the period from start of RPS until the end of premeiotic S-phase taken to be 34 h,[20] stage VII pachytene spermatocytes are 220 − 34 or 186 h beyond this point of reference.

When pairing at zygotene and pachytene is studied with three-dimensional reconstruction techniques, the spermatid stage of development for reference to Oakberg's scheme can be estimated by alternating "LM" histological sections.[21] However, with the commonly used cell suspensions for the whole-mount spreading techniques, stages are mixed up so that only morphological criteria are available. Moses[22] has developed a set of stage criteria for the male based on the behavior of the nucleolus (that shows staining properties analogous to the components of the SCs), the morphology of the attachment points of the SCs to the nuclear envelope, and, most important, the morphology and the amount of pairing between the sex chromosome axes. Other morphological features such as the kinetochores and the degree of spreading of the centric heterochromatin are rather variable and sensitive to the pH of the fixative and, therefore, not easy to adopt. Moses and co-workers[23] have monitored the sequence and the relative durations of the so-defined pachytene substages I through V (see Figure 1) with the aid of a labeling experiment. ^{3}H-thymidine was incorporated at the premeiotic S-phase of the RPS. The sequence of substages established on morphological grounds was found to be correct, as well as the approximate duration of each stage. An alternative way of following spermatocytes through the meiotic stages is to make use of an hydroxyurea-induced front of spermatogonia and spermatocytes, as has been done by Dietrich and de Boer.[24] Hydroxyurea kills germ cells in S-phase by blocking ribonucleotide reductase, thus preventing the accumulation of dNTPs, which inhibits semiconservative DNA replication.[25] Thanks to the fact that the G1 phase of the A1 spermatogonia is of a variable but generally long duration, these cells, at stages VIII and IX of the cycle of the seminiferous epithelium, repopulate the seminiferous tubules after HU-induced killing of the A1 (S + G2) and A2-B spermatogonia and RPSs. They present themselves as a front throughout meiosis that can be harvested on consecutive days, provided enough animals are injected. Since all the cell cycle data of the germinal cells are known, the "meiotic age" of each newly appearing nuclear phenotype can be computed. Generally, the impression of the succession of substages, as established by Moses and co-workers, also emerges from this methodology. Figure 1 compares the relative timing of Moses[22] (based on SCs and nucleoli) with the absolute timing of Dietrich and de Boer[24] (after HU and based on SCs) with the absolute timing of Oakberg,[18,19] based on histological sections and germ cell depletion through radiation. Maximal pairing of the sex chromosomes according to Moses[22] takes place in pachytene substage II, which is positioned slightly earlier than the maximal pairing phase in the HU-front. Using the behavior of the SC as the "true" criterion for zygotene, pachytene, and diplotene, the EM yields a shorter duration (by $\approx$ 2.25 d) of the pachytene stage than the histological method. The amount of pairing between the X and Y chromosomes stands out as the main criterion for within-pachytene age when random samples of spermatocytes are processed for SC analysis.[9,21,22,24,26] The initiation of XY pairing is delayed, compared to the autosomes. After a phase of maximal pairing, that takes 1 d or less, the XY SC is reduced to 10 to 0% of the length of the Y at substage V (Figure 1). Pertinent to the validity of this characteristic is the within-stage variation that, according to Glamann,[9] occurs in CBA inbred males, but has not been found by Dietrich and de Boer[24] in the Swiss random-bred stock. Moses and Poorman[27] noted a decrease in the length of the SCs from early to late pachytene. However, Glamann[9] found exactly the opposite effect, the ratio of mid- to late pachytene total SC length being 104:159. Such an impression can also be gained from the air-dried pachytene spermatocytes on consecutive days of development as presented by Oud et al.[28] and from pachytene morphology in histological sections.[106]

Pairing at zygotene usually starts at the chromosome ends located in each other's vicinity at the nuclear envelope (bouquet stage; see Reference 9). It may occur interstitially as well.[23,24] Guitart et al.[29] used 10 to 17-day-old prepubertal male mice for studying the beginning of meiotic pairing. They also emphasize that pairing can start both interstitially and terminally, but also at more than one location on the bivalent. In addition, they observed that bivalents can be very asynchronous in their pairing behavior, with fully paired and unpaired "bivalents" to be found within the same nucleus.

B. The Female Mouse

In female mice, an analysis of the approximate duration of the meiotic prophase stages could be undertaken by simply sampling meiocytes on the consecutive days of fetal development.[6,27,30] Comparison of the results from these authors is hindered by differences in genetic material and by differences in presentation of the results. No author attempted to assess the absolute duration of the pachytene stage. Thus, no principal answer can be given to the question of whether in the female as in the male, every stage or substage has an absolute duration (as is common with cellular processes) or whether some variation is allowed. Such variation is suggested by the presentation of stages by these authors, usually as their relative frequencies encountered per day of gestation. Another phenomenon complicating the picture and already touched upon in Section II.B, is spontaneous germ cell death, which is relatively rare during the first meiotic prophase in normal males, but occurs abundantly from day 18 of gestation until birth[31] (day of finding the vaginal plug = day 0, C57BL/6J mice) or even commencing 1 day earlier in Swiss (Scofield) albino mice.[17] Both authors locate cell death at pachytene, a stage that subsequently could be "stretched" when judging meiocytes by the appearance and disappearance of the SC alone.[6,30] The following durations of stages are "educated guesses" from the papers cited above: zygotene, 1 to 2 days, commencing at day 14 or 15; pachytene, 3 days, commencing at days 15 and 16 of pregnancy (plug day = day 0). The material cited in this section also seems to be heterogenous with respect to the amount of synchrony, the fetuses of Speed[6] being more regular than those of Dietrich and Mulder.[30] Moses and Poorman[27] comment on a decrease in absolute length of SC per cell from zygotene until the end of pachytene. The initiation of pairing may be different in the primary oocyte, compared to the primary spermatocyte, since the proximal centromeric chromosome ends containing the centric heterochromatin often are unpaired.[6]

IV. CHROMOSOME CONSTITUTIONS THAT LEAD TO A REDUCTION IN FERTILITY

Lists of the diverse types of chromosome mutations which affect gametogenesis quantitatively have been produced by Searle et al.,[14] de Boer and Searle,[32] Searle,[33] and, indirectly, by de Boer.[10] To suit the needs of this chapter, Table 1 has been compiled, which combines elements from Table 1 by de Boer and Searle[32] with the glossary of chromosome mutants treated by de Boer.[10] Only chromosome mutants will be treated here for which enough information concerning the pairing behavior has been collected. Reciprocal translocation heterozygotes will be treated first, followed by Robertsonian translocation heterozygotes, inversion heterozygotes, carriers of more than one chromosome mutation, and aneuploid chromosome mutants. Another very useful source of information about mouse chromosome mutations with and without an effect on germ cell survival can be found in Green,[34] of which currently a new edition is in preparation.

V. CHROMOSOME PAIRING IN RECIPROCAL TRANSLOCATION HETEROZYGOTES

The laboratory mouse has an "all acrocentric" karyotype. The breaks giving rise to reciprocal translocations (abbreviated T) divide the chromosomes involved into two parts: the interstitial segments carrying the centromere plus the proximal chromosome region and the translocated (exchanged) segments containing the distal chromosome region. Thus, the classical reciprocal translocation cross formed through meiotic synapsis of homologous chromosome segments contains four pairing segments, even when it concerns an X-autosomal reciprocal translocation where, in one translocated segment, the Y chromosome is the pairing partner (Figure 2). A special category of translocations are the so-called "three breakpoint rearrangements" or interstitial translocations, for which the term insertion (abbreviated Is) is used by convention. At the level of the SC, the number of investigations devoted to the

Table 1
**A LIST OF MAINLY AUTOSOMAL CHROMOSOME MUTATIONS THAT
PRINCIPALLY OR POSSIBLY LEAD TO A REDUCED GAMETOGENIC
EFFICIENCY**

Chromosomal constitution	Gametogenic effect
Autosome-autosome translocations (some) and an insertion	Variable degree of spermatogenic and spermiogenic impairment
X-autosome translocations	Usual pachytene arrest, but in T16H[a] metaphase I and sometimes metaphase II is reached, while Is1Ct[b] is oligospermic
Y-autosome translocations	Spermiogenic impairment
Heterozygotes for a Robertsonian translocation	Moderate reduction in testis weight and sperm count
Inversion heterozygotes	Unknown
Some multiple translocation heterozygotes without chromosomes in common	Variable spermatogenic and spermiogenic impairment
Heterozygotes for two reciprocal translocations that have one chromosome in common	Variable spermatogenic and spermiogenic impairment
Heterozygotes for two reciprocal translocations that have two chromosomes in common	From absence of an effect to total azoospermia, with cell death starting at pachytene
Heterozygotes for a reciprocal and a Robertsonian translocation that have one chromosome in common	Reduced testis weights and sperm counts, approaching oligospermia
Heterozygotes for two inversions with a chromosome region in common (some)	Variable spermatogenic and spermiogenic impairment
Tertiary monosomics and tertiary trisomics, usually derived from male sterile autosome-autosome translocations	Variable spermatogenic and spermiogenic impairment
Primary trisomy for a small translocation chromosome	Effect comparable to tertiary trisomy
Multiple univalents	Spermiogenic impairment

[a] T(X;16)16H, also known as Searle's translocation.
[b] Is(In7; X)1Ct, an insertion of autosomal material into the X chromosome, also known as Cattanach's translocation.

Extended from De Boer, P. and Searle, A. G., *J. Reprod. Fertil.*, 60, 257, 1980 (Table 1).

pairing behavior of reciprocal and interstitial translocations is only scanty. We will attempt to summarize the data and start with autosome-autosome translocations (A-A), followed by X-autosome translocations (X-A).

A. Autosome-Autosome Translocations

Observations on meiotic pairing primarily are confined to male mice. Initially, these were of a truly qualitative nature, utilizing the EM. In the quantitative studies, pairing analysis was carried out by the LM, which can lead to misinterpretation if the observations are not supported by electron microscopy on the same material.[13,35] Especially when one or two pairing segments are short and an interstitial segment is involved (that carries constitutive centric heterochromatin), a ''centric'' association under the LM can be mistaken for a case of true synapsis with SC formation. Another problem is that with the spreading methods usually employed, the meiotic stage can only be assessed indirectly by the extent of synapsis between the sex chromosomes (see Section III.A).

The qualitative studies of Choi[8] (by three-dimensional reconstruction of fetal oocytes) and Moses et al.[3] (by whole-mount spreading) have yielded the insight that in reciprocal translocations for which the male is fertile, complete synapsis can be achieved. Both authors compared the lengths of the segments in the translocation cross with the lengths of the same segments in G- and Q-banded mitotic chromosomes. The agreement between the two as-

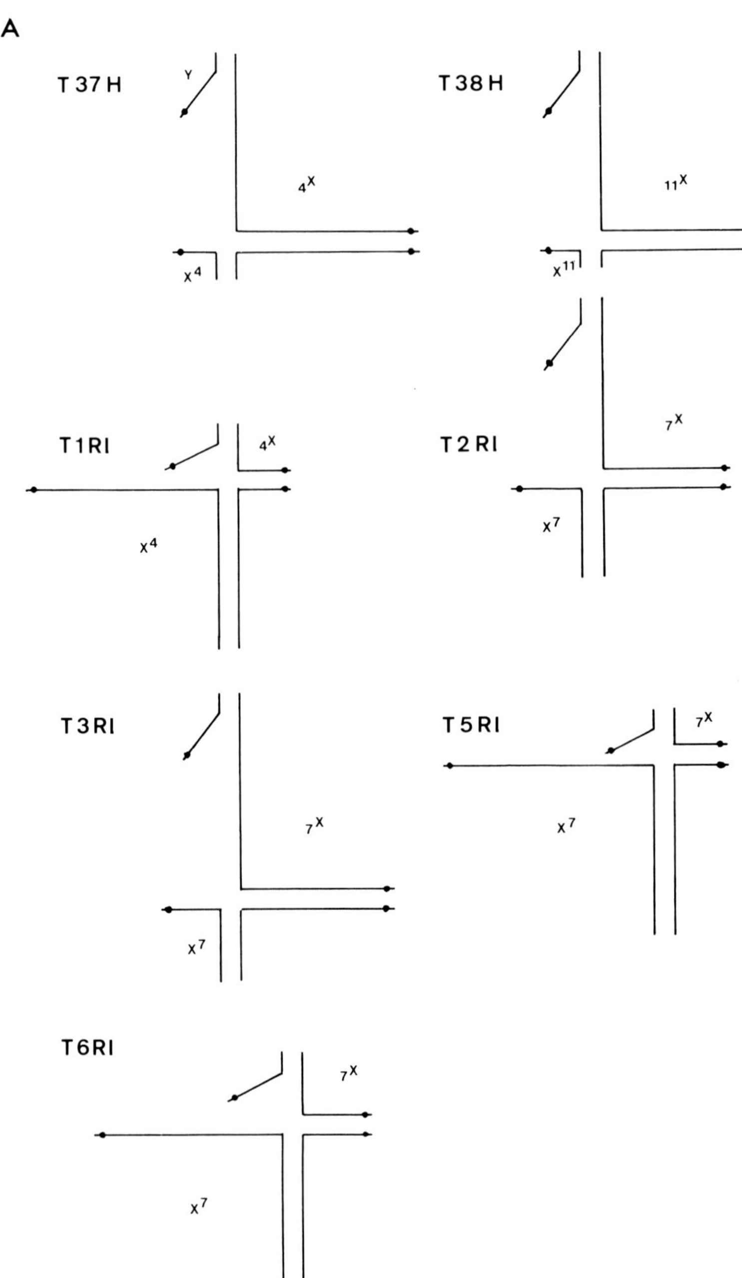

FIGURE 2.   Pachytene pairing diagrams of most of the chromosome mutants treated in this text. (A) X-autosome translocations, (B) autosome-autosome translocations, (C) double heterozygotes for two semi-identical translocations, (D) trisomic karyotypes. Centromere positions are indicated by a dot.

sessments (assuming only homologous pairing around the center of the cross) is as good as one can expect, given the limited amount of material.

The quantitative studies by Forejt et al.[36] and de Boer et al.[13] are entirely devoted to the comparison of male-fertile and male-sterile reciprocal translocations. Figures 3A, B show two rings of four multivalents of T70H, B with nonhomologous meiotic pairing around the center of the "cross". Table 2 gives the data regarding multivalent composition and vicinity

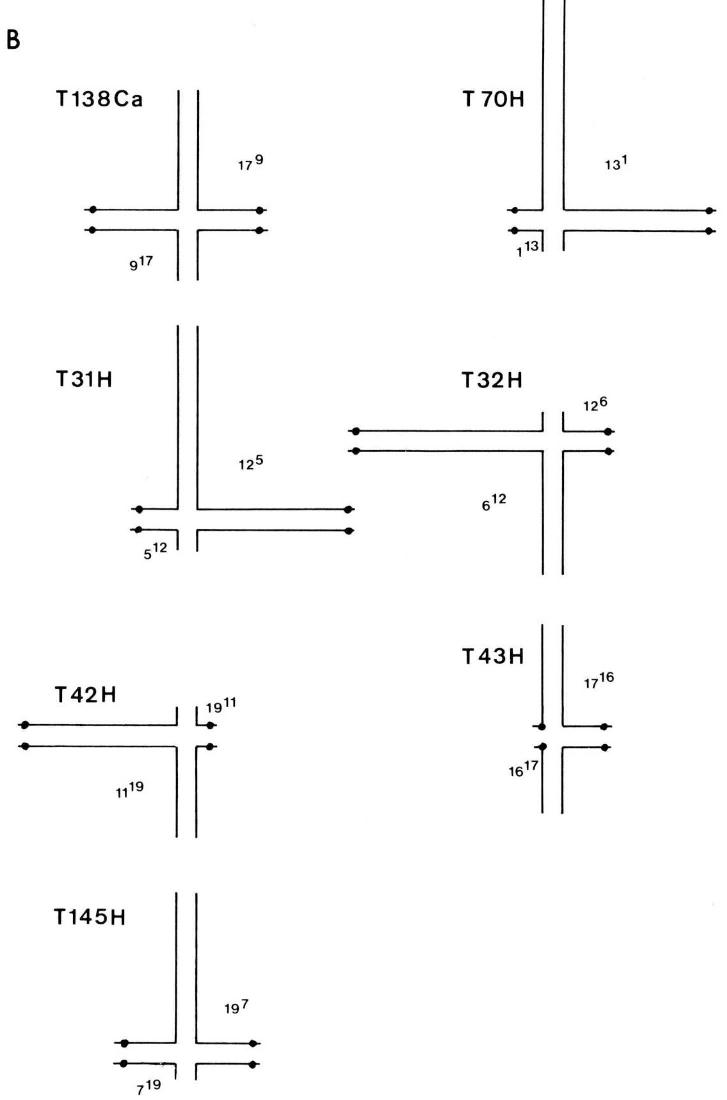

FIGURE 2B.

of the multivalent to the sex bivalent. Forejt et al.[36] included a male-fertile reciprocal translocation control that had both breakpoints about halfway along the chromosomes involved. De Boer et al.[13] used a control reciprocal translocation in which breakpoints were more like the male steriles, with always one proximally and one distally located. All reciprocal translocation pairing crosses are given in Figure 2B. When pooling their data, de Boer et al.[13] stated that the amount of "contact" between the multivalent and the sex chromosomes in the spread preparations strongly depended on the availability of unpaired chromosome segments for interaction, the univalent of the CIII + I configuration being most frequently involved (90.6%), with CIV involved 73.2% of the time, and RIV being lower (27.8%) (C = chain, R = ring). Figure 4A shows an example of "contact" for a CIV of T70H. When a similar exercise was undertaken in the translocations studied by Forejt et al.,[36] the data were RIV, 28.3% and CIV and others, 60.8%. Forejt et al.[36] divided spermatocytes into two groups, one with less than 10% of the Y chromosome paired (with the X) and another with more of it paired, the first fraction representing all but the first 2 to 3 d of pachytene.

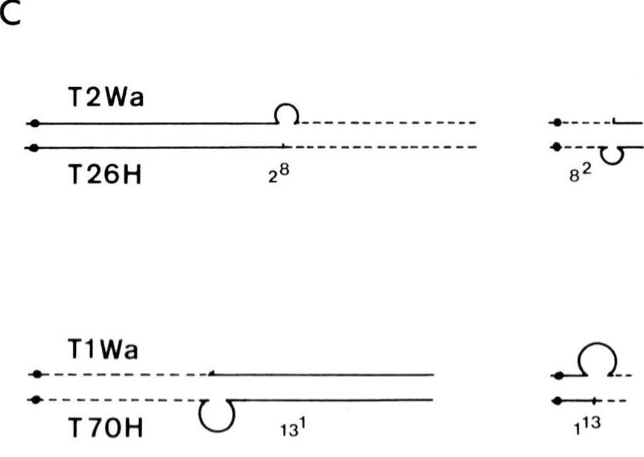

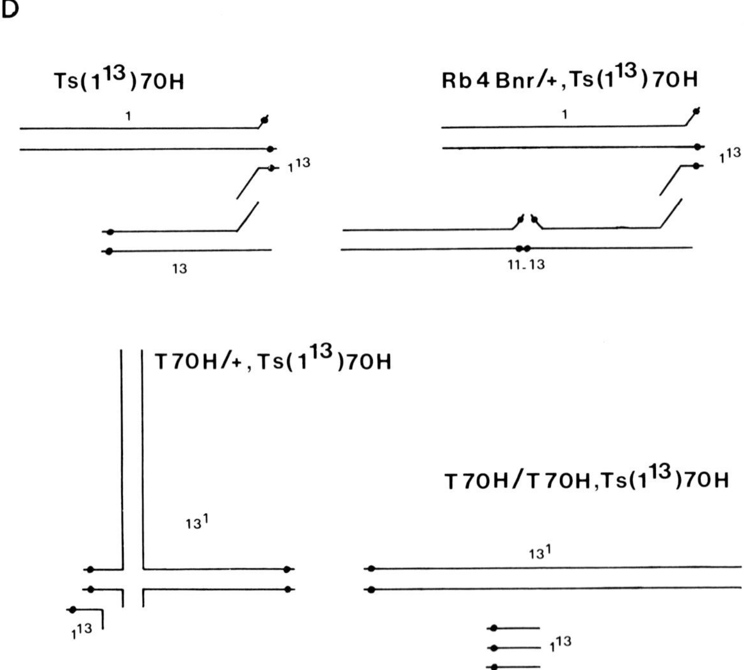

FIGURE 2C and D.

They did not find different ratios for the multivalent types, except for T138Ca/ + , which had CIVs more abundant in the later pachytene stages. De Boer et al.,[13] when pooling all the reciprocal translocations studied, observed increases for RIV and CIII + I, with a concomitant decrease for CIV when meiosis proceeded. Looking again at their data, this effect primarily was caused by the T32H/ + and T70H/ + karyotypes and not so much by T31H/ + and T42H/ + . The aspect of stability of pairing will receive further attention in Section X.B.

The distributions of the arbitrarily defined X-Y pairing classes were strikingly alike for the different translocations both in the material of de Boer et al.[13] and in the material of Forejt et al.[36] ($X_6^2 = 5.54$ and $X_3^2 = 5.43$, respectively). In the former material, there was

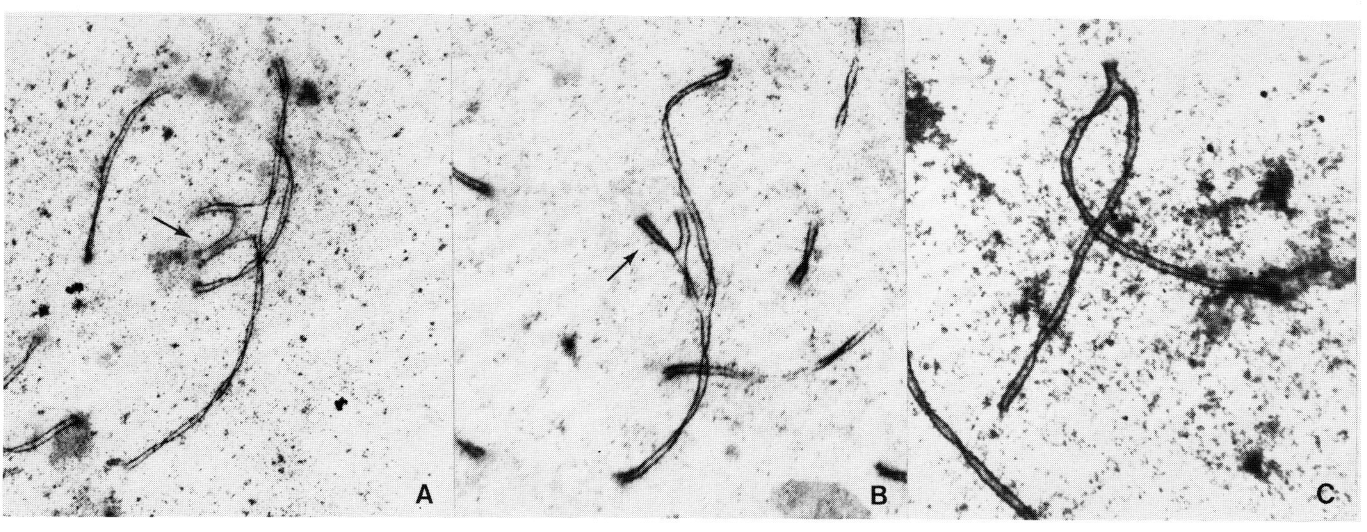

FIGURE 3.   Electron micrographs of silver-stained spread nuclei of the T(1;13)70H/ + ring quadrivalent (A and B) and the Rb(11.13)4Bnr/ + trivalent (C.) The T70H short interstitial segment is indicated by an arrow.

Table 2
MULTIVALENT SPECTRA AT PACHYTENE OF MALE-FERTILE AND MALE-STERILE RECIPROCAL TRANSLOCATIONS IN THE MALE

Translocation	Relative sperm count (%)	No. cells	Configuration (%)[a]				% XY contact[b]	Ref.
			RIV	CIV	CIII + I	CIV and others		
T(16;17)43H	3	216	18.5			81.5	56.9	36
T(5;12)31H	8.9	113	27.4			72.6	61.1	36
T(7;19)145H	<3	101	18.8			81.2	64.4	36
T(9;17)138H	>85	106	81.1			18.9	8.5	36
T(5;12)31H	8.9	101	44.6	21.8	33.7		60.4	13
T(6;12)32H	5.3	94	36.2	50.0	13.8		50.0	13
T(11;19)42H	0	81	14.8	70.4	14.8		87.6	13
T(1;13)70H	80.9	99	82.8	16.2	5.0		30.3	13

Note: Reference 36 used more than one male while in Reference 13, always two males were sampled. For T31H/+ and T70H/+ these two males were heterogeneous.

[a] R = Ring; C = Chain.
[b] Contact usually refers to apposition of unpaired autosomal segments and the axial elements of the sex chromosomes.

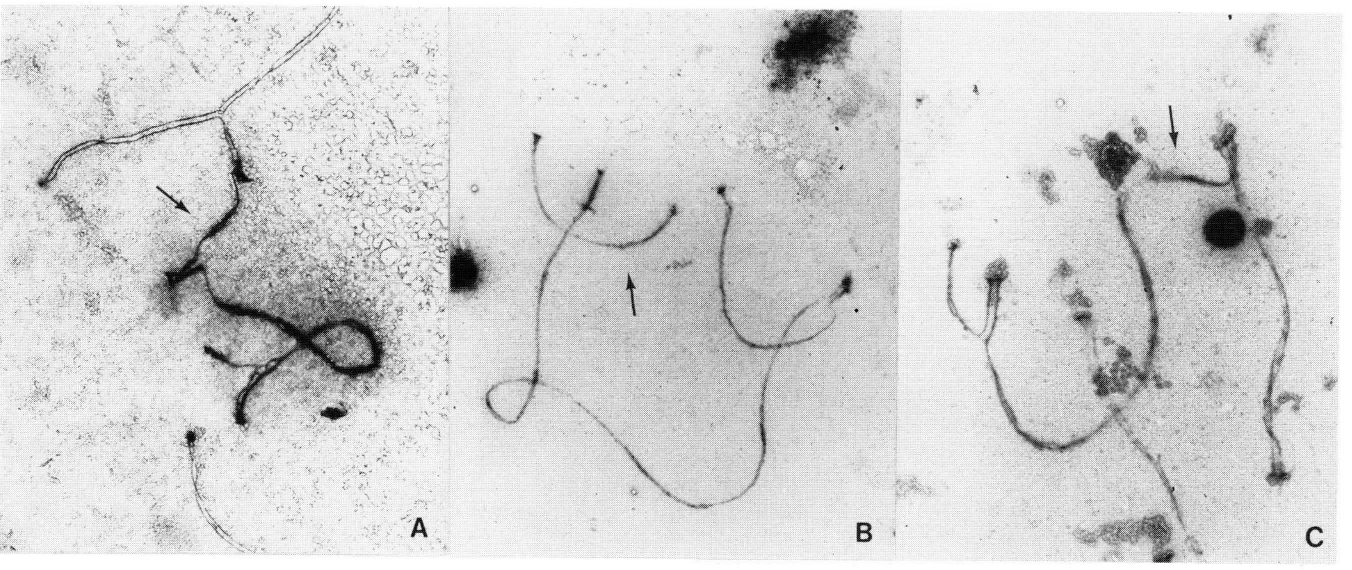

FIGURE 4. (A) Contact between chr 1^{13} (arrow) of the T70H/+ chain quadrivalent and the X chromosome. The pairing situation of the distal end of chr 1^{13} with its homologous segment on chr 13 is not clear. (B, C) Inclusion of the axial element of chr 1^{13} (arrow) in the sex vesicle of Ts(1^{13})70H male mice. Chr 1^{13} is univalent in B, but part of a trivalent in C. The spermatocyte C is at pachytene substage II; the one of B, substage V (see Figure 1).

no trend for more late pachytene cells in the fertile control (T70H/ +), which was the case in the latter (T138Ca/ +).

The data of Table 2 have been used to compute Spearman rank correlation coefficients between the sperm count and the percentage of ring (RIV) configurations (r_s = 0.89, p <0.01), between the sperm count and the frequency of "contact" between the translocation-involved chromosomes and the sex vesicle (r_s = −0.89, p <0.01), and between the frequency of ring configurations and the proximity of the translocation-involved chromosomes to the sex vesicle (r_s = 0.86, p <0.05). From these computations, it can be concluded that unpaired segments or chromosomes are preferably found close to sex vesicle chromatin. Both parameters (i.e., "closeness" and lack of pairing) are highly negatively correlated with the number of sperm.

Table 3 compares the multivalent spectra for two male-sterile and two male-fertile reciprocal translocations between pachytene and diakinesis-metaphase I. The conclusions to be drawn from this Table are that lack of chiasma formation in an interstitial segment does not lead to spermatocyte, spermatid abortion (T70H/ +). Chiasma formation primarily seems to be a function of length when it concerns an interstitial segment (compare T70H/ + with T138Ca/ +). Moreover, for the male sterile translocations, it is clear that pairing in the short interstitial segment is not very effective with regard to chiasma formation, especially if one takes into account that the diakinesis-metaphase I sample is a selection of the pachytene population. In T31H/ + , for instance, cell death commences at pachytene stages IV to XI, though with considerable interanimal variation,[13] which is also expressed by the occasional occurrence of oligospermic male fertiles,[37]

The pachytene pairing behavior of male-sterile reciprocal translocation has never been studied in female meiosis. However, the initial belief that male-sterile translocations had no effect on gametogenesis in the female proved to be incorrect. In female T42H/ + mice, the count of dictyate oocytes on days 3-5 after parturition is only 35% of normal,[38] which explains why at this age the gonads are smaller.[39]

The only translocation in which a male/female comparison has been made is the insertion Is(7;1)40H.[40] Male SCs were also studied by Searle et al.,[15] who provided the basic information about this very interesting rearrangement. Is40H is remarkable in that pachytene spermatocytes die abruptly at stage IV of the cycle of the seminiferous epithelium, in comparison to most male-sterile chromosome mutations where death occurs over a wide range of stages from pachytene to spermiogenesis, with probably most cell mortality at metaphase I (see References 10 and 13). In the females, gonads are smaller (first postnatal week,[41]) which probably is explained by a lower number of dictyate oocytes (see T42H/ + above). Consequently, the reproductive lifespan of Is40H females most likely is reduced.[15] Table 4 gives the multivalent spectra for pachytene oocytes and spermatocytes and for diakinesis oocytes after the *in vitro* culture of adult dictyate stages liberated from antral follicles. Male and female pachytene configuration spectra are very different (X_1^2 = 28.7 p <0.005). Contrary to what could be expected from the generally higher chiasma frequencies in female mice,[42] chromosome pairing in mid-pachytene oocytes is less complete than during the first half of pachytene in spermatocytes (see Figure 1 for a relative position of Oakberg's stage IV, when Is40H spermatocytes die). The diakinesis figures of Is40H females do not seem to represent the pachytene configurations in that many bivalent-carrying meiocytes have been lost (X_1^2 = 22.2, p <0.005). Thus the selection against meiocytes with two heteromorphic bivalents with interstitially unpaired segments most likely operates at the end of pachytene and onset of diplotene (see Section III.B).

Another investigation into male pachytene pairing of Is40H has been carried out by Searle et al.[15] Their data are not totally comparable with those of Mahadevaiah et al.[40] since the subdivision into configuration types was different. It was shown that Is40H can also pair as a reciprocal translocation (RIV and CIV). In these cases, the proximal segment of chr 1[7]

Table 3
COMPARISONS OF MULTIVALENT SPECTRA AT PACHYTENE AND AT DIAKINESIS-METAPHASE I FOR TWO MALE-FERTILE AND TWO MALE-STERILE RECIPROCAL TRANSLOCATIONS IN THE MALE

Karyotype	Sterile/fertile	Meiotic stage	Number		Configuration (%)					Ref.
			Males	Cells	RIV	CIV	CIII + I	II + II	II + I + I	
T(7;19)145H/ +	Sterile	Pachytene	>1	101	18.8	81.2[b]				36
		Diakinesis	5	534	15.4	58.0	18.9	7.7		103
T(5;12)31H/ +	Sterile[a]	Pachytene	>1	113	27.4	72.6[b]				36
			2	101	44.6	21.8	33.7			13
		Diakinesis	3	221	4.1	39.8	54.8	1.3		37
T(1;13)70H/ +	Fertile	Pachytene	2	99	82.8	16.2	5.0			13
		Diakinesis	15	540	2.2	74.3	22.4	1.1		42
T(9;17)138H/ +	Fertile	Pachytene	>1	106	81.1	18.9[b]				36
		Diakinesis	10	1537	65.6	23.3			11.2	103

[a] An occasional oligospermic fertile.
[b] CIV and others.

Table 4
MULTIVALENT CONFIGURATIONS OF MEIOCYTES FROM Is(7;1)40H AT PACHYTENE (MALE AND FEMALE) AND DIAKINESIS (FEMALE)

		Number		Configuration (%)			
Sex	Meiotic stage	Animals	Cells	MIV[a]	CIII + I	II + II	Ref.
♀	Diakinesis	>1	107	71.1	3.7	25.2	15
♀	Pachytene	5	168	46.4		53.6	40
♂	Pachytene	4	176	74.4		25.6	40

[a] M = multivalent.

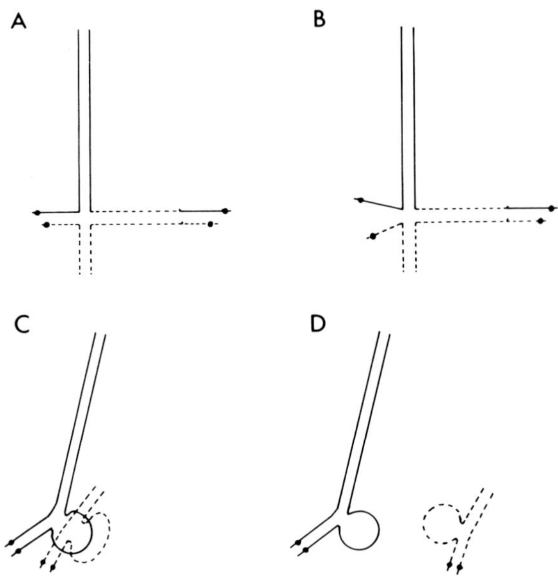

FIGURE 5.   Pachytene pairing diagrams of Is(7;1)40H. A and B depict nonhomologous pairing after homologous pairing initiation from the inserted segment. In C, there is predominantly homologous pairing, whereas, in D, there is lack of pairing for the inserted segment.

pairs nonhomologously with the proximal segment of 7, while with RIV this is also the case between the same two segments but located in the chrs 1 and 7⁻ (see Figure 5). Both publications comment on the position of the bivalents/quadrivalents which was frequently close to the sex vesicle. However, contact does not seem to be as intense as in, for instance, T42H/+ (see Table 2). Mahadevaiah et al.[40] scored 56.7% positive contact and Searle et al.[15] gave 39.6% as a figure. In both male and female Is40H bivalents with a loop, pairing adjustment (a phenomenon we will encounter when treating paracentric inversion heterozygotes, Section VII) was most likely the mechanism which led to apparently normal SCs in 6.1% of oocytes and 2.2% of spermatocytes. The hairpin loop shown by Mahadevaiah et al.[40] most likely is a transitional stage of this process.

Since female primary oocytes are, to some extent, asynchronous, Mahadevaiah et al.[40] were able to substage their pachytene oocytes in early, mid- and late pachytene. Most surprisingly, the fraction of bivalent-carrying oocytes was found to rise from 43.6% (early and mid-pachytene pooled) to 66.7% in late pachytene ($X^2_1 = 7.76$, $p < 0.01$). Dissociation of the lateral elements of the SC in the loop region by absence of a cross over is not excluded as the explanation of this phenomenon.

The fact that early pachytene still was collected on days 18 and 19 of pregnancy (counting plug day as day 0) points in the direction of meiotic delay with Is40H females.

B. X-Autosome Translocations

Meiotic pairing of X-A translocations has mainly been studied by Ashley.[35,43-46] Reciprocal translocations between the X and an autosome (surprisingly, the chrs 7 and, to a lesser extent, 4 are overrepresented[107]) almost by definition cause sterility in the male. As a parallel to the A-A translocations, the study of the meiotic pairing behavior of these translocations could throw a light on a possible origin of this sterility. Breakpoints, as with male-sterile translocations, tend to cluster in the proximal and distal region of the X, while the autosomal breakpoint usually is near a chromosome end also. Thus, marker chromosomes can be produced.[47] These X-A translocations cause spermatocyte death at pachytene. When breakpoints are both halfway along the X and the autosome, as with T(X;16)16H, first meiotic division figures can be seen.

Table 5 contains the multivalent spectra (and Figure 2A, pachytene pairing diagrams) of seven X-A translocations at pachytene, two of which (T37H/+ and T38H/+) were also studied at diakinesis in the female. A comparison of these two translocations suggests that T37H/+ pairs more efficiently in the short translocated autosomal segment, being $\approx$ 1.75 times as long as its T38H/+ counterpart. (By analogy to A-A translocations we assume short interstitial segments to be rather devoid of chiasmata.) A similar tendency is reflected by the male pachytene data. Thus, there is more meiotic pairing in T37H/+. In agreement with the above given rationale (Section V.A.), testis weights are higher in T37H/+ males than in T38H/+ males. Also, as judged by the amount of XY synapsis found in the pachytene samples studied, spermatocyte death, on average, occurs later in T37H (where an occasional diakinesis cell can be seen) than in T38H/+ (no meiotic figures visible[47]).

In the material studied by Ashley, two translocations (R2R1 and R3R1) follow the behavior of T37H and T38H, but in an even more distinct way since the translocated autosomal segments are longer (see Figure 2A): CIV configurations are exclusively seen.

When the interstitial autosomal segment is short, however, a different picture emerges. For these segments, there is either asynapsis or desynapsis after initial synapsis, both routes leading to a considerable frequency of heteromorphic bivalents (see Table 5). The shorter the interstitial autosomal segment (R5), the higher the bivalent frequency. When studying the R1 series of X-A translocations, Ashley and Russell[46] discovered the phenomenon of nonhomologous synapsis around the center of the (hypothetical) pachytene pairing cross (Figure 2A). This seemed to originate from one translocation breakpoint close to or within a prominent Giemsa-positive band (most breakpoints are in Giemsa-negative bands). Giemsa-negative and Giemsa-positive chromosome bands are known to be related to the chromomere pattern of pachytene bivalents (the chromomeres relating to Giemsa-positive staining regions produced after appropriate treatment of mitotic chromosomes; see Reference 48). Giemsa-positive bands replicate during the second half of S phase and contain the ontogenic genome, contrary to the Giemsa-negative bands, which have a preponderance of housekeeping genes.

Pairing between the sex chromosomes is almost normal in X-A translocations, especially when the breakpoint in the X chromosome is proximal (T37H, T38H, T2R1, and T3R1). Y univalents (the II + I + I configurations of Table 5) also are hardly found in T1R1, T5R1, and T6R1, which is remarkable because of the shortness of the translocated X segment. There are even indications of a more persistent synapsis of the Y in these situations.[44,45] Another observation of Ashley pertains to nonhomologous synapsis in heteromorphic bivalents, which is more frequently encountered when nonhomologous synapsis also occurs in the quadrivalents. Thus, X and autosomal material can be nonhomologously synapsed.[45]

Table 5
MULTIVALENT SPECTRA OF X-AUTOSOMAL TRANSLOCATIONS IN THE MALE (PACHYTENE) AND FEMALE (DIAKINESIS)

Karyotype	Sex	Meiotic stage	No. cells	Configuration (%)					Ref.
				CIV	CIII + I	II + II	II + I + I		
T(X;4)37H/ +	♀	Diakinesis	51	76.5	21.5	2.0			47
	♂	Pachytene	100	87.0	13.0				
T(X;11)38H/ +	♀	Diakinesis	55	60.0	40.0				47
	♂	Pachytene	100	31.0	69.0				
T(X;4)1Rl	♂	Pachytene	74	64.9	1.4	32.3	1.4		46
T(X;7)2Rl	♂	Pachytene	25	100					43
T(X;7)3Rl	♂	Pachytene	19	100					35
T(X;7)5Rl	♂	Pachytene	101	20.8	3.0	71.3	4.9		35
T(X;7)6Rl	♂	Pachytene	76	42.0	4.0	53.0	1.0		43

Pertinent to the purpose of this chapter is what connections can be drawn between the pairing behavior thus observed, the pachytene stages at which they occur, and the subsequent fate of the pachytene spermatocytes. Unfortunately, no attempt at meiotic staging has been undertaken. Neither histological data nor testes weights of these X-A translocations has been supplied.

X-A translocations follow the pattern set by A-A translocations in that one "insertion" has been studied (Is[In7; X]1Ct or Cattanach's translocation) Figure 2A. A further parallel is that, in both cases, chr 7 donated the segment and that the segment donated to the X in Is1Ct(7C − [7E3-F1]) largely is congruent with the segment donated to chromosome 1 in the case of Is(7;1)40H (7B1 — 7F1).

Also comparable with Is40H, Is1Ct forms quadrivalents (pairing in the inserted segment, 38%) and bivalents (62%, n = 44, EM observations). LM observations on a much larger sample (n = 516) yield proportions of 28% quadrivalents and 72% bivalents.[45] These figures do not differ very much from the 20% pachytene quadrivalents seen in air-dried preparations of pachytene spermatocytes by Slizynski (1967, cited by Eicher[49]), of which 5 to 10% still persist at diakinesis-metaphase I.[49] In agreement with the smaller size of the insertion and perhaps helped by the facts that the proximal autosomal breakpoint is in a Giemsa-positive band and that the insertion is surrounded by X-chromatin which has a different chromatin configuration in male meiosis (see Chapter 3, this volume), the bivalent frequency is higher in Is1Ct than in Is40H (compare with Table 4, males). However, desynapsis of an initially synapsed "insertion" (compare with Is40H females) could also lead to the higher bivalent fraction in Is1Ct. Remarkably, Is1Ct is the most fertile X-A translocation known, with an epididymal sperm count 21% of normal,[50] while Is40H was the most sterile A-A translocation. Pairing adjustment leading to a seemingly normal 7;7⁻ bivalent, with the 7 lateral element meandering around the 7⁻ lateral element, was seen in the EM in 10% of Is1Ct male nuclei.[45] A systematic cytological study of pairing and chiasma position in Is1Ct (including careful staging of the pachytene nuclei) could give an answer to the question of desynapsis by an absence of crossing over in situations of partner exchange.

VI. CHROMOSOME PAIRING IN HETEROZYGOTES FOR ROBERTSONIAN TRANSLOCATIONS

Despite the vast number of Robertsonian translocations (Rb) from the wild (feral) and the growing number of *de novo* Robertsonian translocations that arise in mouse laboratory stocks carrying feral Robertsonian translocations, no printed systematic study of chromosome pairing of Robertsonian trivalents through zygotene, pachytene, and diplotene is available. The observations published have been fragmentary. Most of them agree with the findings of Moses et al.[51] about heterozygous Robertsonian systems in Lemur hybrids. Thus, the proximal segments of the acrocentric chromosomes with centric constitutive heterochromatin are the last to pair (see Figure 3 of de Boer, Reference 10). Compared with the normal bivalents, complete synapsis of the trivalent is delayed.[52,53] Often, the two nonhomologous acrocentric proximal regions can be seen to pair (or associate) nonhomologously. No positive evidence for a central element has been reported (see Figure 3C). More prominent in late cells, the side arm of the Rb SC can have disappeared, resulting in a continuously running SC.[108] Both this phenomenon and the nonhomologous pairing of the proximal segments of the acrocentric chromosomes could be taken as expressions of pairing adjustment (see Section VII). Brown and Burtenshaw[52] noted that, in 7.5% of 200 pachytene spermatocytes of Rb(2.18)6Rma, there was contact during pachytene between one unpaired proximal segment of an Rb-involved acrocentric chromosome and the nonpairing segment of one of the sex chromosomes (usually the X). A similar low frequency of sex vesicle contact was observed in the Rb(11.13)4Bnr and Rb(9.19)163H males studied by de Boer.[10] A special Rb translocation is Rb(X.2)2Ad. Of the 52 cells studied by EM whole-mount spreading, only 2 exhibited complete pairing of chr 2. This led to infrequent association of the 2 and Y centromeric regions at late pachytene.[54]

Generally, Robertsonian heterozygosity (for a single translocation) leads to reduced testis weights and reduced sperm counts and a slight increase in the frequency of malformed spermatozoa.[10,55] However, fertilization rates are not affected.

VII. CHROMOSOME PAIRING IN INVERSION HETEROZYGOTES

No quantitative data are available on the efficiency of spermatogenesis in male paracentric inversion heterozygotes (In/ +). Chromosome pairing in these rearrangements will only be briefly commented upon in the framework of this chapter. For more information, the reader is referred to References 10, 27, and 56. Thus, at early pachytene the noninverted proximal and distal chromosome ends show delayed meiotic pairing and a classical inversion loop is produced when the inversion does not span most of the chromosome. If the inversion spans most of the chromosome, a loop is not produced, but instead proximal and distal ends will pair nonhomologously, although delayed. In the "delayed phase", contact between the unpaired axes of the "inversion construct" and the sex chromosomes is established. When, during pachytene substages II to IV (Figure 1), "synaptic adjustment" ensues (the inversion loop disappears by a system of desynapsis inside the loop[22,56]), unpaired chromosome ends are no longer seen and X, Y contact is thus alleviated. Fertility problems usually are not encountered with In/ + males, but this only indicates at the least a 7.5 to 10% production of normal, wild type spermatozoa (see Section II.A). In females, much the same sequence of events is observed.[27]

Just by accident, Burgoyne and Baker[17] discovered a significant reduction by 35% in the number of oocytes of In(X)1H females shortly after birth. Tease and Fisher[57] have now studied chromosome pairing in this large inversion, spanning most of the X chromosome. The inverted segment was often found to be dominant for the initiation of meiotic pairing, the proximal and distal segments showing delayed pairing. Thus, inversion loops were found in 28.7% of pachytenes at day 15, gradually sinking to 10% at day 18 and 7.6% at day 19 of pregnancy (plug day = day 0). These authors, in agreement with Moses and Poorman,[27] attributed this effect to "synaptic adjustment". Some of the loops were double, suggesting incomplete pairing. As on the later days of gestation the frequency of this configuration did not exceed 2% of the total number of oocytes, it cannot be solely responsible for the drop in surviving oocytes, as determined by Burgoyne and Baker.[17]

One sterile male carrying a pericentric inversion of Rb(11.13)4Bnr has been studied at pachytene by EM.[58] The inversion changed the metacentric Rb4Bnr chromosome (see Figure 3C) into a submetacentric one which, in this male, paired with acrocentrics 11 and 13. Configurations very much resembled the Rb(11.13)4Bnr trivalent, apart from the presence of a loop. Again, the proximal parts of chromosomes 11 and 13 were found to exhibit delayed synapsis and an "appreciable amount" of contact with the X, Y axial elements was established. The early pachytene stages seemed to be overrepresented among the nuclei analyzed. Subsequently, as with paracentric inversions, the loop could be resolved by "synaptic adjustment". Among the meiotic stages in air-dried preparations on glass, there was a shortage of metaphase I primary spermatocytes, of metaphase II secondary spermatocytes (more severely), and very few mature sperm heads were seen. In metaphase I cells, there appeared to be an appreciable amount of X, Y univalents (63.3%, n = 30).

VIII. CHROMOSOME PAIRING IN COMBINATIONS OF CHROMOSOME MUTATIONS

The general finding in mice with more than one chromosome mutation is that the quality and quantity of the spermatogenic process is impaired.[10] This author also gives a relatively complete treatise on the fertility data so far collected in male mice with more than one chromosome abnormality. Here we will confine ourselves to the limited data on mutants, where observations on meiotic chromosome pairing have been made. These observations concern male mice with more than two different Robertsonian translocations with and without chromosome arms in common (monobrachial homology)[53,59] and male mice that carry two reciprocal translocations between the same two chromosomes, of which two different karyotypes were studied.[13] As these studies were more quantitative, we will give the results first. Table 6 shows the main results of five T(1;13)70H/T(1;13)1Wa and two T(2;8)26H/T(2;8)2Wa male mice. Figure 2C contains pachytene pairing diagrams. Both sets of reciprocal translocations have breakpoints so little different that multivalent pairing usually is not achieved and never is stable. Hence, for each karyotype, the meiotic pairing behavior of a long and a short heteromorphic bivalent with an interstitial lack of homology can be studied (Figure 6). Theoretically, both heteromorphic bivalents should show a buckle which, at the first half of pachytene, disappears through "synaptic adjustment" during pachytene substages II to IV if their behavior is analogous to that of the duplication heterozygote Dp(7)1Rl and the inversion heterozygotes (Section VII) studied by Moses and Poorman.[22,56,60]

In the 13^1 and 2^8 long marker bivalents, "adjustment" apparently takes place very readily, as no clear pairing buckles (or loops) were ever seen. Infrequently, the longer lateral element was seen to meander around the shorter one at the site of the translocation breakpoints. The 1^{13} and 8^2 short marker bivalents were very interesting in their behavior, however. When using the amount of X, Y pairing as the indicator of meiotic stage, no "pairing adjustment" was seen to take place. However, pairing was often complete and involved an increase in the length of the shorter one of the two short marker chromosomes (1^{13}70H and $8^2$2Wa, respectively). Frequently, in these cases, the bivalent had a horseshoe shape, with the lateral element of the longer "homolog" clearly thickened (Figure 6C). In agreement with expectation, "contact" with the axial elements of the sex bivalent was more intimate in the short marker bivalent, when pairing failed at the chromosome ends rather than interstitially (configurations with a buckle, see Figure 6D, E). The frequency of "contact" of the heteromorphic bivalent with the sex bivalent could be correlated either with "unpaired autosomal regions" as the one variable, or with the sperm count as the other variable. The former correlation was higher ($r_s = -0.96$, $p < 0.01$ for the combination of karyotypes) than the latter ($r_s = -0.63$, N.S.). Chiasma formation has been studied in both the long and short marker bivalents of T1Wa/T70H.[61] In 521 diakinesis-metaphase I primary spermatocytes from three fertile T70H/T1Wa males, 1.7% contained univalents, a percentage not far from

Table 6

FERTILITY ESTIMATES OF THE T(1;13)70H/T(1;13)1Wa AND T(2;8)26H/T(2;8)2Wa KARYOTYPES IN THE MALE AND MEIOTIC PAIRING BEHAVIOR OF THE SMALL HETEROMORPHIC BIVALENT IN PACHYTENE SPERMATOCYTES

		Karyotype			
		T(1;13)70H/T(1;13)1Wa		T(2;8)26H/T(2;8)2Wa	
% Fertile males[a]		36.0	(N = 25)	79.4	(N = 34)
Sperm count (% normal)		13.7	(N = 25)		
Aberrant epididymal sperm heads		41.3 ± 27.0	(N = 13)		
$N_{males}/N_{meiocytes}$		5/233		2/100[b]	
% Meiotic pairing configurations		60.5	(27.0)	43.0	(2.3)
(% "positive" sex vesicle con-		19.3	(57.4)	28.0	(17.9)
tact in parentheses)		16.5	(80.0)	26.0	
		3.7	(77.8)	3.0	(65.5)

Note: See also Figure 2C.

[a] A male is declared fertile here when at least one conceptus, expressed as a decidual reaction, is produced after two matings.

[b] The two males used here were very different. One had a very high sperm count (high-normal), while the other was virtually azoospermic. This hampers the comparison between the two karyotypes.

From De Boer, P., Searle, A. G., van der Hoeven, F. A., de Rooij, D. G., and Beechey, C. V., *Chromosoma,* 93, 326, 1986. With permission.

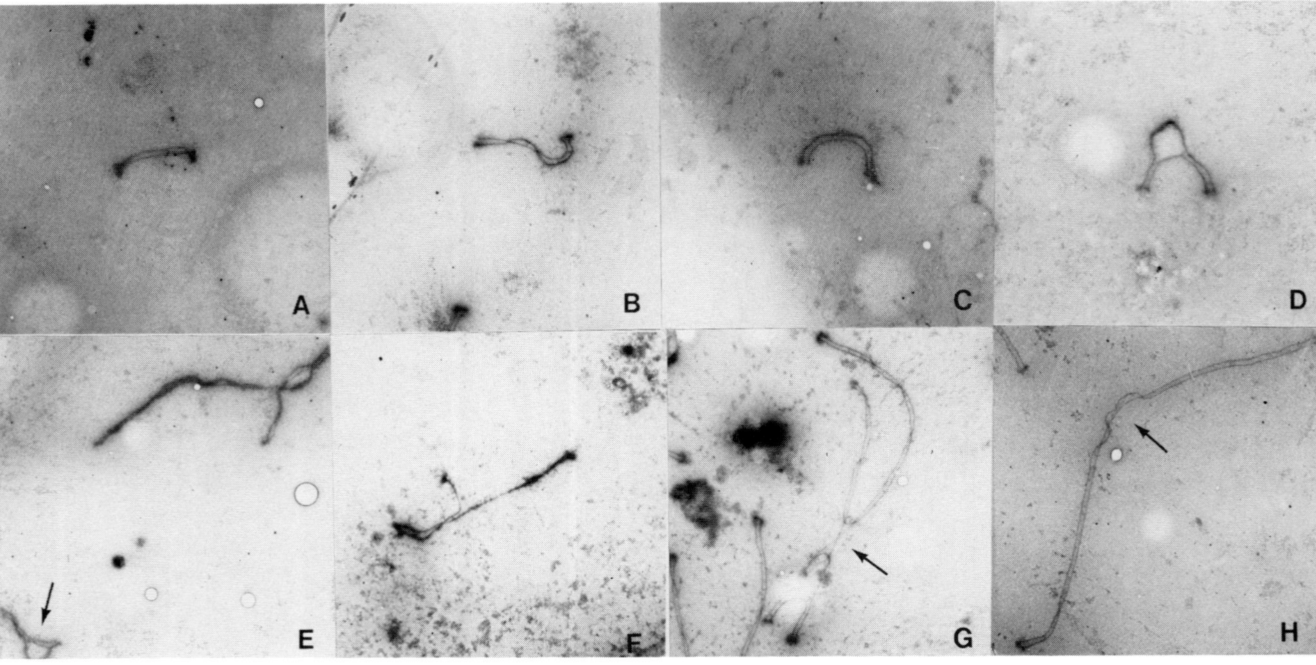

FIGURE 6.   Electron photomicrographs of the translocation-involved marker bivalents of T26H/T2Wa (B, D, F, G) and T70H/T1Wa (A, C, E, H). (A-F) The small marker bivalents 1^{13} and 8^2 are shown with decreasing levels of pairing adjustment. (B, C) The longer lateral element is bolder. (D—F) Thickening of unpaired lateral elements is shown. (E) A pseudosynaptonemal complex is formed (arrow). (G) The homologous loops of the marker bivalents attempt to form a synaptonemal complex (arrow). (H) The long marker bivalent SC is shown, which, apart from its length, in this case can be picked up by wrinkles around the site of SC adjustment (arrow). (From De Boer, P., Searle, A. G., van der Hoeven, F. A., de Rooij, D. G., and Beechey, C. V., *Chromosoma*, 93, 326, 1986. With permission.)

Table 7
MULTIVALENT SPECTRA OF MALE AND FEMALE TERTIARY TRISOMICS

| Karyotype | Sex | Meiotic stage | Number | | Configuration (%)[a] | | |
			Animals	Cells	Univalent (I)	Trivalent (III)	Ref.
Ts(1[13])70H	♂	Pachytene	1	41	73.7	26.3	66
	♂	Diakinesis	5	501	79.0	21.0	42
	♀	Diakinesis	23	218	85.3	14.7	42
Ts(5[12])31H	♂	Pachytene	3	62	96.8	3.2	67
	♀	Pachytene	3	40	80.0	20.0[b]	67

[a] In both karyotypes, complete synapsis by means of nonhomologous pairing in a trivalent is only rarely observed.

[b] It is not clear whether there is homologous synapsis.

the 0.8% in T70H/T70H males. Also, the long marker bivalents 13[I]T70;13[I]T1Wa and 13[I]T70;13[I]T70H had identical chiasma frequencies. (The T1Wa long marker chromosome is somewhat shorter than the T70H long marker chromosome; see Figure 2C).

Gropp et al.[62] gave an account of the fertility characteristics of multiple heterozygotes for Robertsonian translocations. Male mice double heterozygous for two Robertsonian translocations with monobrachial homology form chain quadrivalents at the first meiotic metaphase. From the data collected by Gropp and co-workers, 36% of the 42 different constructs studied were sterile, whereas a further 29% showed morphological changes in the testes. In the sterile double heterozygote Rb(16;17)7Bnr/Rb(8;17)1Iem, spermatogenesis drops to a low level during diplotene — the early spermatid stages.[62,63] In double Robertsonian heterozygotes with monobrachial homology, chains can also be seen at pachytene.[53] The proximal regions of the acrocentric chromosomes may remain unsynapsed and show a thickened axial element and frequent contact with the axial elements of the sex bivalent. These phenomena do seem to be only more severe in longer chains. With five or more participating chromosomes (≥ three metacentrics and two acrocentrics), sterility always follows. Contrary to these complex chains, rings consisting of alternating metacentrics can involve up to 16 chromosomes without truncating spermatogenesis.

The three-dimensional reconstruction of such rings has been briefly described by Wolf and Winking.[64] Nuclear architecture was profoundly changed and three to four clumps of heterochromatin intermingled with euchromatin were found in the center of the pachytene nucleus, rather than the eight to ten blocks at the periphery in the "all acrocentric" karyotype. The efficiency of the meiotic divisions is reduced.[62,63]

An example of pachytene pairing in a sterile male that carried two overlapping paracentric inversions of chromosome 1 has been documented by Chandley.[65] Of the 150 cells analyzed, 90% showed a loop which, in the great majority of cases, was situated adjacent to the sex vesicle.

IX. CHROMOSOME PAIRING IN ANEUPLOID MOUSE MUTANTS

The karyotypes to be covered in this section are two tertiary trisomics (Ts[5[12]]31H and Ts[1[13]]70H), one translocation trisomic of T70H origin, and a primary trisomic, also of T70H origin. Pachytene pairing diagrams are given in Figure 2D.

All data available on meiotic pairing and chiasma formation in tertiary trisomic males and females are given in Table 7. Although, in the male late pachytene stages, the frequencies of sex vesicle contact of the small translocation chromosomes are very high (close to

100%[66,67], Ts(1^{13})70H has a relatively good spermatogenic efficiency (Table 8). Total sterility, however, (the status of Ts[5^{12}]31H), can be induced by a change of genetic background.[66] In female meiosis of Ts(5^{12})31H, the small marker chromosome 5^{12} more frequently associates with "other bivalents" than in males.[67] Its axial element, as in male tertiary trisomics, always is thickened. No indications of homologous pairing are given (Table 7). Also, trivalent formation at diakinesis is lower in Ts(1^{13})70H females than in males (Table 7).

In 3 to 5-day-old Ts(5^{12})31H females, the number of oocytes is reduced by 71%, compared to normal litter mates.[68] This reduction is high enough to lead to a 18% reduction in litter size.[37] The numbers of oocytes from antral follicles of adult Ts(1^{13})70H females are significantly decreased,[16] and we have the impression of a reduction in litter size and reproductive lifespan also. Primary trisomics for the small 1^{13}T70H marker chromosome resemble Ts(1^{13})70H in a number of ways. The sperm count is slightly lower and the frequency of abnormal spermatozoa is slightly decreased (Table 8). Table 9 shows the pachytene pairing data of one adult male and five female fetuses. Figure 7A — E illustrates the meiotic pairing possibilities of chr 1^{13}. Complete homologous synapsis by SC formation between the three 1^{13} small marker chromosomes (Figure 7A) seems to be more frequent in the female than in the male.

All fertility problems associated with the meiotic pairing aspects of the extra chromosome seem to be alleviated when chr 1^{13} has improved chances for complete homologous pairing, as in T70H translocation trisomics. Abnormal spermatozoa are no longer encountered (Table 8) and degenerative metaphase I meiotic figures, which are so abundant in tertiary and primary trisomics, are virtually eliminated. Table 10 gives the spectra of multivalent configurations at male pachytene and diakinesis. Figure 8 shows the three dominant multivalent configurations, CIII + II, CV, and CIV + I, at male pachytene. T70H translocation trisomy is the only karyotype that shows a chimeric behavior with regard to univalence and "XY contact" (always correlated). Mainly the CIV + I configuration (and, to a lesser extent, chains of V that have the interstitial segment of chr 1 unpaired; see Figure 8B) show this type of involvement.[69] CIV + I diakinesis cells are remarkable in that they exhibit precocious chromosome spiralization and a delayed passage to the second meiotic division.[69,70] Speed and de Boer[70] explain why these meiocytes unexpectedly are rescued and undergo an essentially normal spermiogenesis.

When viewing the pachytene pairing data and the chiasma data of the three trisomic karyotypes in the male, the coincidence between the frequencies of pachytene univalence and late diplotene, diakinesis univalence for the small marker chromosome is striking, despite the hardly quantitative approach of the pachytene pairing studies. These results clearly show that the length of a SC in the distal telomeric region can be extremely short and yet allow crossing over. They also demonstrate that in situations of presumed partner exchange, as in the CV of translocation trisomics and CIII of primary trisomics, the crossing over frequency per unit length can be increased over the normal bivalent situation, where usually one chiasma is observed in the 1^{13} bivalent (male and female T70H homozygotes[71]).

X. DISCUSSION

Three aspects of male and female germ cell development in carriers of chromosome mutations that lead to a reduced efficiency of the gametogenic process will be discussed here, ordered from more general to more specific: (1) the kinetics of first meiotic prophase, (2) the relationship between meiotic pairing and germ cell death, and (3) why is the fertility depression more severely felt in the male than in the female and why are qualitative disturbances during spermiogenesis more the rule than the exception?

Table 8

SPERM COUNTS AND MORPHOLOGY (FROM THE CAPUT EPIDIDYMUS) AND FREQUENCIES OF "CONTACT" BETWEEN TRANSLOCATION CHROMOSOMES AND THE SEX CHROMOSOMES DURING PACHYTENE OF T70H TRISOMIC KARYOTYPES

Karyotype	No. males	Sperm count	% of +/+ control	% of trisomic control[a]	% of abnormal sperm heads (No. males/	(No. cells)	% of sex vesicle contact (No. males/	No. cells)	Ref.
T70H/+,Ts(1^{13})70H	22	257 ± 88	.50.2	100	5.0	(7/700)	{ 9 { 20[b]	(1/100)	70,69
T70H/T70H,Ts(1^{13})70H	15	79 ± 59	15.4	24.9	63.3	(15/1500)	83	(1/53)	[c]
Ts(1^{13})70H	22	109 ± 44	21.3	34.4	77.2[d]	(9/1350)	≈100	(2/102)	70,66
T70H/+	32	414 ± 111	80.9		4.3	(7/1000)	30.3	(2/99)	79,13
+/+	55	512 ± 125	100		9.3	(7/1000)	Low		79,13

[a] As no indication for germinal cell death was found and spermatozoal morphology was normal, the reduction in sperm output of T70H translocation trisomics is blamed on the trisomic state per se.

[b] 9% inclusion of a univalent chromosome 1^{13} in the sex vesicle.

[c] Data mainly of W. Veldman, Agricultural University internal report.

[d] Morphology of uterine spermatozoa.

Table 9
MULTIVALENT SPECTRA OF MALE AND FEMALE T70H
"PRIMARY" TRISOMICS

Meiotic stage	Sex	Number		Configuration (%)		
		Animals	Cells	III	III[a]	II + I
Pachytene[b]	♂	1	58	13—21	4—12	75
Diakinesis[c]	♂	15	577	14.7		85.3
Pachytene[d]	♀	5	100	47	4	49

[a] The three small marker chromosomes associate, but there is doubt as to the occurrence of a central element between one largely unpaired marker chromosome and the bivalent (see Figure 7C, D).

[b] Whole-mount spreads, inspected by EM.

[c] Only concerns stages 1 and 2 as defined previously[104] (excluding the contracted bivalents). Data from W. Veldman, Agricultural University internal report.

[d] Whole-mount spreads, inspected by LM. Data from H. Savelkoul, Agricultural University internal report.

A. Disturbances in Meiotic Rate

In the male sterile construct Rb(16.17)Bnr/Rb(8.17)/Iem (see Section VIII), a delay of 18 h was shown for the first late diplotene cells after labeling at the premeiotic S-phase some 10.5 d earlier.[72] Moreover, the heterogeneity in the kinetics of those meiocytes that reached diakinesis was greater. This was the main tendency in the work of Speed and de Boer,[73] where, in the oligospermic mutants Rb4Bnr/ + and Ts(1¹³)70H, the front of meiocytes was not delayed, but the average delay for the remaining cells correlated negatively with the sperm counts and positively with the frequency of degenerating meiotic figures at diakinesis metaphase I. T70H translocation trisomics form a special category since degenerative metaphase I figures cannot be seen and spermatozoal morphology is normal (see Table 8). Yet there is variation in the length of the period from premeiotic S-phase to diakinesis as well. This has been blamed on the CIV + I and CIII + II configurations, the former characterized by a univalent and the latter by nonhomologous pairing between the proximal segment of chr 1 and the distal segment of chr 13 (see Figure 8). Despite the fact that in T70H/ + no increased heterogeneity in the length of first meiotic prophase was found, CIII + I configurations reached diakinesis systematically later than CIV configurations.[73] This again could be an indication of an effect of a lack of meiotic pairing and subsequent crossing over on meiotic rate.

The question is which (sub)stage of meiotic prophase is most sensitive to meiotic delay? Here the evidence is only indirect, as a pachytene pairing analysis has never been undertaken with the aim to study meiotic prophase kinetics. It is assumed that in the mouse crossing over takes place at the beginning of pachytene (substage II of pachytene according to Moses[22], Figure 1) when the number of ^{3}H thymidine grains of the P-DNA synthesis associated with the SC is greatest.[23] The total number of recombination nodules and bars per nucleus at this stage is in good agreement with a normal chiasma count, and recombination bars are seen for the first time.[9] It could be argued that delays just before and at early pachytene (when Z-DNA synthesis is taken to be preparatory for chromosome pairing and subsequent crossing over[74]) have a disturbing effect on meiotic pairing and crossing over. A chromosome mutant capable of illustrating meiotic pairing problems by chiasma frequency at diakinesis is a combination of Rb(11.13)4Bnr and Ts(1¹³)70H (see Figure 2D). Telomeric pairing of the 1¹³T70H small marker chromosome now has to compete with chr 13 in a situation of Robertsonian heterozygosity. De Boer and Nijhoff[75] found the association frequency of 1¹³ in healthy looking primary spermatocytes up to diakinesis to be increased from 21% (n =

Table 10
MULTIVALENT SPECTRA OF T70H TRANSLOCATION TRISOMIC MALES

	Number		Configuration (%)			
Meiotic stage	Males	Cells	CIII + II	MV[a]	M IV + I[a]	Ref.
Pachytene	1	28	46	36	14	69
Late diplotene-early diakinesis	6	132	62.9	24.2	12.9	69

[a] M = multivalent.

501, the tertiary trisomic) to 33% (n = 122, X_1^2 = 7.04, p <0.01). Apparently, the fact that one of the 13s is a partner in a Robertsonian translocation creates more opportunity for chromosome 1[13] to pair at the distal telomeric end. Nevertheless, spermatogenesis is severely affected (Table 11) and chromosome morphology at the meiotic divisions is poor.

An experimental meiotic prophase delay in T70H/+ mice of 25 h at diakinesis was achieved by Wauben-Penris et al.[76] when they restricted a population of surviving resting primary spermatocytes to a developmental period of 5 to 10 h. The chiasma frequency was especially increased in the short interstitial segment that experiences the greatest pairing difficulties (see Figures 2B and 4A). In this situation, spermatocyte morphology was entirely normal. The conclusions to be drawn from these experiments could be that (1) meiotic prophase delay, as such, is not deleterious for meiocyte development and (2) meiotic pairing and crossing over, normally time restricted, receive a better chance in the mutation-involved chromosomes when the duration of meiotic prophase is increased.

Among Is40H primary oocytes, Mahadevaiah et al.[40] observed early pachytene cells on days 18 and 19 of development (day of plug = day 0), which seems to be rather late (compare with Section III.B). A delay in prophase I has also been observed by Burgoyne and Baker[17] in XO female fetuses. However, this could be fully accounted for by the general developmental delay of 4 h observed in this karyotype.

Cell physiological aspects of developmental delay have hardly been studied. Erasmus and Grootegoed[109] followed rRNA processing in pachytene spermatocytes of T70H/T1Wa and +/+ males. Contrary to somatic cells, germinal cells showed a 32S ribosomal RNA fraction as a clearly visible intermediate step between processing from 45S to 28S, the 32S/28S ratio being 0.56 for controls and 0.90 for the male sterile mutant. Thus, processing from 32S to 28S seems to be delayed. The same investigators also found lower ATP levels (25% of the controls) in male sterile T70H/T1Wa pachytene spermatocytes.

B. Comparisons of Chromosome Pairing Between Males and Females

In a few mutants, meiotic pairing has been studied in both male and female meiosis: Is40H (see Table 4), Ts(5[12])31H (see Table 7), and primary trisomics for the T70H small translocation chromosome (see Table 9). No conclusive picture emerged from these studies. Judged from the synaptic capabilities of extra small marker chromosomes, female prophase might be more pairing competent, especially in the case of the primary trisomics. The more extensive study of Is40H shows that, for the interstitial initiation of pairing in chr 7, the male has a greater potential.

Even if we take into account that aneuploidy per se may lead to a reduction of gamete production*, a reduction of gamete output accompanies the meiotic pairing problems encountered in both sexes. Moreover, in the normal "all acrocentric" mouse karyotype, univalence, for both autosomes and sex chromosomes (see section X.C.3) has been related to sterility. Winking and Johannisson[78] briefly described asynapsis or partial pairing of

* See Table 8, this Chapter and the observation that a male mouse tetrasomic for the T31H5[12] small marker chromosome was sterile despite consequent meiotic pairing between the two extra bodies.[77]

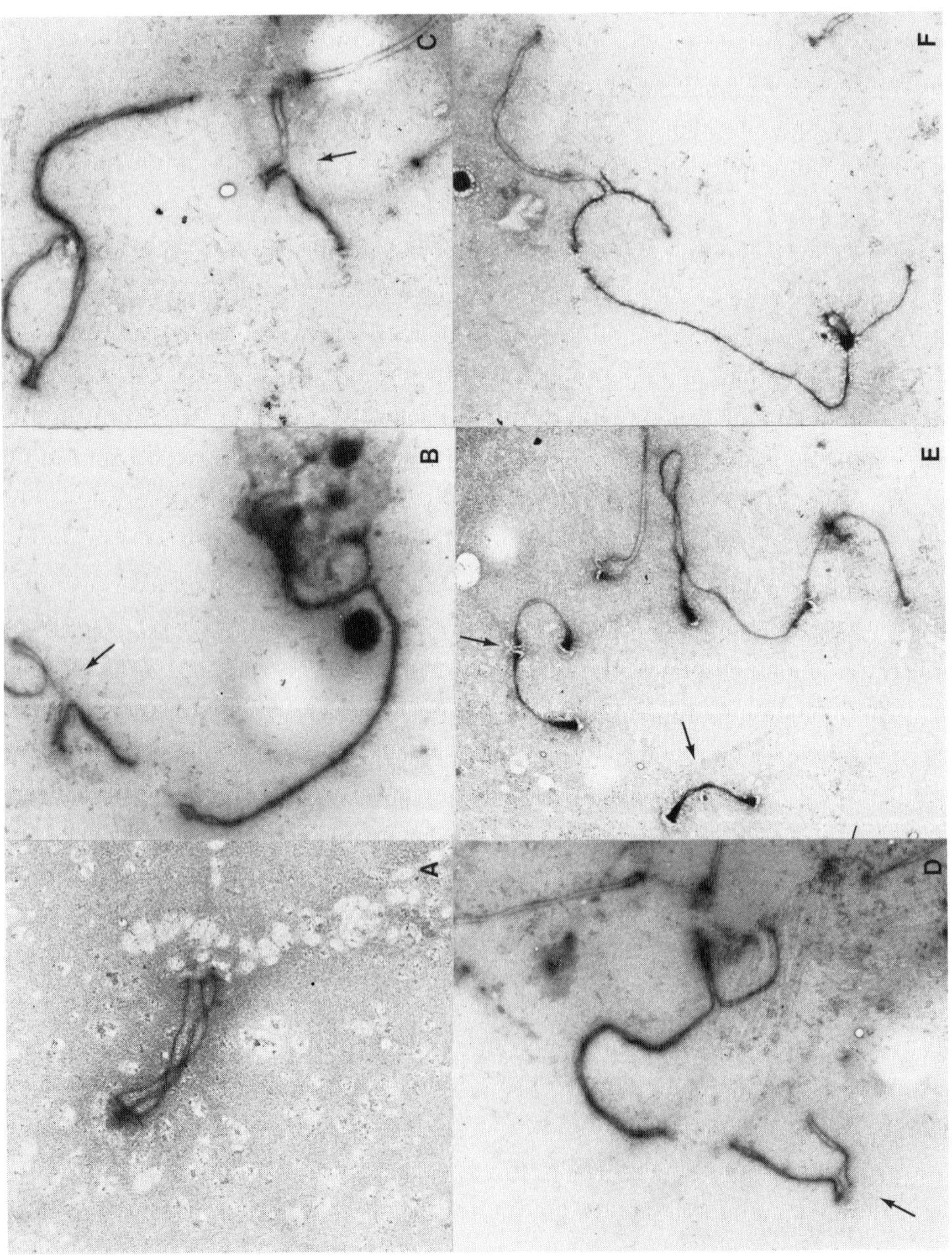

FIGURE 7.   Electronmicrographs of pachytene spermatocytes of T70H/T70H, Ts(1^{13})70H. (A). Full homologous pairing of the three 1^{13} small marker chromosomes. (B) Classical partner exchange within a 1^{13} trivalent. However, the 1^{13} close to the X is thickened. (C, D) Partial pairing of one chr 1^{13} (also see Table 9). (E) Desynapsis at late pachytene, early diplotene. Note that, in the cases depicted (B, C, D), the thickened 1^{13} axial element is longer than the synapsed lateral elements. Arrows, chr 1^{13}. (F) Contact between a partially asynaptic bivalent and the sex chromosomes in a normal $+/+$ mouse.

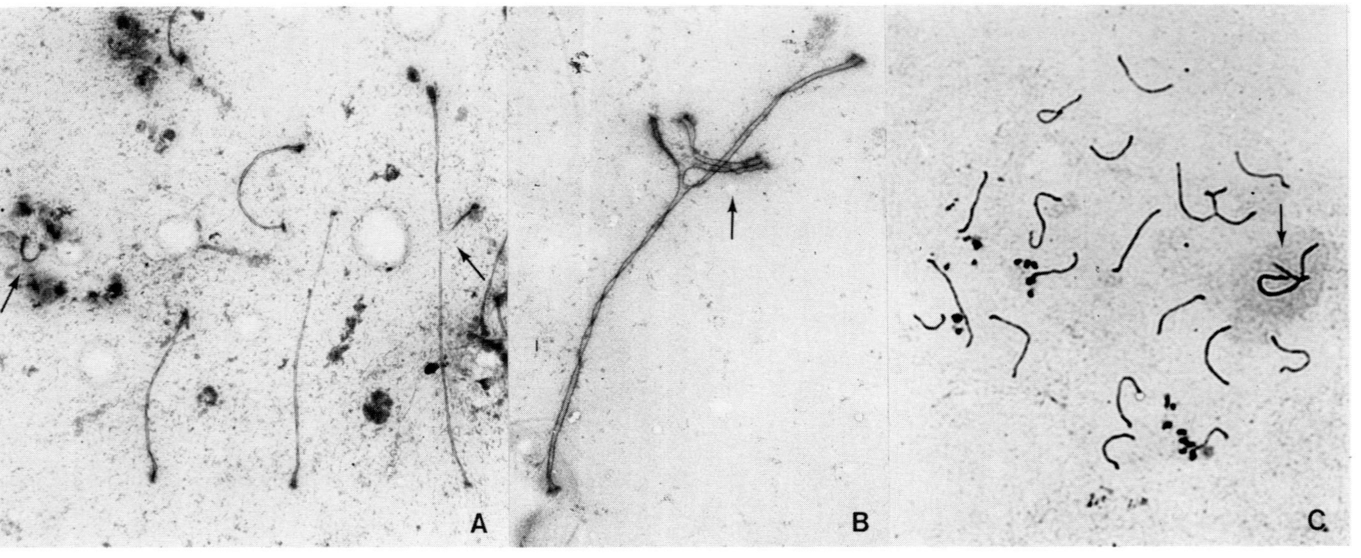

FIGURE 8.   The predominantly occurring multivalent frequencies of T70H/+, Ts(1^{13})70H. (A) Trivalent plus bivalent pairing, both arrowed, the trivalent one pointing at a nonhomologous meiotic pairing situation. (B) A chain of V with chr 1 unpaired for the interstitial segment. The distal half of chr 1^{13} shows triple pairing (arrow). (C) A chain of IV plus a univalent associating with the sex chromosomes in the sex vesicle (arrow) (LM silver staining). (From De Boer, P. and van Beek, M. E. A. B., *Chromosoma*, 87, 303, 1982. With permission.)

Table 11
RELATIVE SPERM COUNTS (CAPUT EPIDIDYMIS) AS OBSERVED AND EXPECTED (UNDER THE ASSUMPTION OF INDEPENDENCE), IN CARRIERS OF TWO CHROMOSOME MUTATIONS

| | | (%)[a] | | |
Karyotype	No. males	Observed	Expected	Ref.
T(1;13)70H, + / + ,T(2;8)26H	18	59.0	80.9	105
T(1;13)70H, + / + ,Rb(11.13)4Bnr	14	16.8	34.1	10
Rb(11.13)4Bnr/ + ,Ts(1¹³)70H	6	7.9	9.0	75

[a] All percentages are relative to + / + controls.

autosomes in sterile male hybrids between *Mus musculus musculus* and *M. musculus domesticus* (both with an "all acrocentric" karyotype). Figure 7F shows that also in this situation, the unpaired segments of the autosomes and sex chromosomes seek contact. Another argument for meiotic pairing problems as a source of a reduced gametogenic efficiency is the decrease in sperm output observed after combining various chromosome abnormalities in one carrier. Table 11 lists some examples where the sperm counts of all the components were known. The decrease in spermatogenic efficiency is equal or greater than that predicted on the basis of independence. In other words, it appears that more than one pairing puzzle poses extra problems for the meiocyte. An identical conclusion can be drawn from the study of long (four or more chromosomes) chains due to Robertsonian heterozygosity[62] (Section VIII). A demonstration of this principle has not yet been produced in the female.

One question to be raised after having looked at the array of mouse mutants presented here is whether homologous meiotic pairing must be followed by a cross over in order to remain stable. The few cells, analyzed in female prophase of Is40H would argue for such a mechanism (see Section V.A). It is generally known that in mouse reciprocal translocations, when there are two adjacent short pairing segments, one interstitial and one translocated, chiasma formation is much higher in the short translocated segment.[79] Yet RIV formation during pachytene must be the rule in these translocations, if "male fertile" (see Table 3). Hence, in short segments relatively rich in constitutive centric heterochromatin, a cross over is not necessary to stabilize meiotic pairing during pachytene. It seems likely that when cross over formation fails in both short segments (e.g., T70H/ + , see Section V.A), the "unlinked" chromosome becomes displaced. This effect could be enhanced in X-A translocations with short (autosomal) interstitial segments (see Table 5 and References 35 and 43).

There is evidence that nonhomologous autosomal pairing can protect against cell death through lack of meiotic pairing. In a way, the nonhomologous pairing of centric heterochromatic regions in the Rb/ + karyotype may be regarded as an example of this, although, because of the uniform nature and localization of mouse satellite DNA around all autosomal centromeres, not a typical one. A more illustrative example of the efficiency of nonhomologous pairing is seen in trivalents of reciprocal translocations, such as the III + II configurations of the T70H translocation trisomics (see Figure 8A and Section IX). Here, nonhomologous pairing takes place between the centromeric and telomeric ends of the two nontranslocation chromosomes. Spermatogenesis is essentially normal. This mode of pairing must be the mechanism by which T31H monosomics (lacking the 5¹² small marker chromosome) have a higher sperm count (7% of normal) than T31H tertiary trisomics (0.4% of normal), despite their overall poorer phenotype.[37] It is tempting to speculate that nonhomologous meiotic pairing is the mechanism by which the adjacent two originating "unbalanced" viable karyotypes 10, 10, 10[13] and 13 of T199H and 16, 17[16], 17, and 17 of T43H are "male fertile" (the latter karyotype with a sperm count of 46% of the controls[10,80]).

Nonhomologous pairing seems to be the mechanism by which the T70H/T1Wa and T26H/T2Wa karyotypes treated in Section VIII escape total sterility (the former) or even reach up to normal fertility (the latter). Here, we obtained evidence that nonhomologous pairing can be operative from the first stages of meiotic prophase. This conclusion is strengthened by Ashley and Russell,[46] when they relate nonhomologous pairing around the center of a reciprocal translocation pairing cross to the positions of one of the two breakpoints in or close to a prominent G-dark band (see Section V.B). Such a position apparently is stimulating for nonhomologous pairing around the translocation breakpoints. Prior to these discoveries, pairing characteristics of mammalian meiosis were thought to be stage constricted: exclusively homologous during zygotene and pachytene substage I and nonhomologous during stages II to IV by "synaptic adjustment" (see Section VII, the so-called two phase pairing theory of meiosis formulated by Rasmussen[81] and subsequently illustrated in the mouse by Moses and co-workers[22,56,60]).

When axial elements are unpaired, many authors observe them to be thicker than lateral elements in normal tripartite SCs. The same "doublings" that can be seen in the axial elements of the sex chromosomes also take place in the autosomal ones.[13] These authors also observed lateral elements of unequal thickness in an SC, suggesting that crossing over is not prohibited by a structural asymmetry of this type. However, it seemed as if a less pronounced "thickening reaction" was positively correlated with the sperm count (T26H/T2Wa). In line with this observation, unpaired "lateral" elements that are observed in situations of partner exchange never thicken. These situations, which are the most frequent occurrence in the large ring multivalents that can be brought about by multiple Robertsonian translocation heterozygosity, do not lead to a dramatic reduction in the spermatozoal counts.[63]

The effects of a lack of meiotic pairing on germ cell fate are thought to be rather independent of the region(s) of the genome it concerns. One exception to this rule could be constructed from the observations presented in this chapter when the two largely identical insertions Is(7;1)40H and Is(In7; X)1Ct are compared. The former, unlike all other male-sterile chromosome syndromes, leads to a truncation of spermatogenesis at stage IV of the cycle of the seminiferous epithelium (see Figure 1). The latter, however, is the mildest of the X-autosome rearrangements, with appreciable numbers of spermatozoa and occasional fertility. As their distal breakpoints in chr 7 are approximately the same, this effect could be due to the segment 7B1-7C harboring a gene (or genes) that is (are) dependent on meiotic pairing for proper functioning. However, it could equally well be argued that the location of the "unpaired" chr 7 insertion is the prime cause for the contrast. In Is1Ct, the location is X-chromosomal in an area which normally never pairs. One might suppose that some of the tolerance of this area to absence of pairing is conveyed to the autosomal segment.

C. Sex-Specific Aspects of Meiotic Pairing Problems
1. Events During Pachytene

When male and female pachytene are compared, the longer duration in the male is very conspicuous (between 6 and 7 d vs. approximately 3 d; see Section III). Nevertheless, phases of meiotic pairing observed in the male also take place in the female and are most likely similar. For instance, "pairing adjustment" involved in the disappearance of the paracentric inversion loop takes place in female meiosis, identical to what is observed in male pachytene substages II to IV; nonhomologous meiotic pairing also is a reality in female meiosis.[27] In the first half of male pachytene and in female pachytene, hnRNA synthesis is relatively low.[31,82] In agreement with this, chromatin structure is rather compact during approximately the first half of male pachytene and relaxes in the second phase.[9,28] Histologically, pachytene spermatocytes can be seen to grow during stage IV of the cycle of the seminiferous epithelium.[110] At this stage, female pachytene is released and cells enter prediffuse diplotene, where transcription increases. In the male, the increase in hnRNA synthesis occurs during the second half of pachytene[82] and thereafter decreases in the postpachytene phases to diakinesis-metaphase I. During the second phase of male pachytene, a longer total SC length in three-dimensional reconstructions can be measured. In line with this, the axial elements of the sex vesicle grow,[9] which is also apparent in many whole-mount spreads (see Figure 4B). Nevertheless, ^{3}H uridine incorporation cannot[82] or can barely[83] be traced in sex vesicle chromatin. At the end of pachytene, the attachment points of the lateral elements of the SC to the nuclear envelope become deltoid and lateral (now partly axial) elements remain visible at the beginning of diplotene.[24] Again, such deltoid outgrowths are not observed in female SCs, nor are the axial elements preserved postpachytene.

In the second phase of male pachytene, there is transport of silver-staining material (Ag-NOR) from the mouse satellite bivalents to one location on the periphery of the sex vesicle, where a conspicuous tripartite body is formed, whose center is "silver-positive."[84] It is often seen in the whole-mount spreads used to study SCs. Thus, when female pachytene cells are in a phase of massive cell degeneration, male pachytene cells become metabolically very active, do not experience excessive cell death,[85] extend their pachytene life for at least another 3 d, and move on to the meiotic divisions. We propose this phase to be connected with male-specific functions, such as spermiogenesis. Also, the foundation of the often observed functional sperm heterogeneity in mammals could find its origin here.

2. Spermiogenesis-Specific Genes

Obviously, many genes do play a role during spermiogenesis, and the ones that are transcribed exclusively post meiosis are best known (examples are *protamin,* located on chr 16, and the oncogene *int-1* on chr 15; for a complete list and references, see Reference 86). However, it is quite plausible (see above) that some of the RNA synthesis during the male-specific pachytene stages also relates to spermiogenesis and sperm function. An influence of the source of the Y chromosome on sperm morphology has been noted.[87,88] Moreover, sperm motility aspects have tentatively been allocated to the Y chromosome (see Reference 89) and fertilization behavior also is under a Y chromosomal influence.[90] Thus, some of these genes must be under control of Y-specific sequences if the structural loci are not Y linked. One possible candidate has recently been isolated:[91] a DNA probe from a Y-enriched library exclusively hybridized on Northern blots from testis RNA.

One of the mysteries of male meiosis is how the Y chromosome exerts this influence. Probably, the ^{3}H uridine incorporation method has not been applied sufficiently to the expanded sex vesicles that can be observed during "male-specific" pachytene. Early replicating bands are found in the pairing region of the Y chromosome, while another one is just below the centromere.[92] The results of *in situ* nick translation experiments on sex vesicles and diakinesis sex bivalents seem to show DNase I-sensitive "active" chromatin on the same locations,[93] although positive signals are not always obtained.[94] Without the isolation of the Y chromosomal DNA sequences of concern, the situation seems to be difficult to resolve.

3. Meiotic Pairing of the Sex Chromosomes

Sex chromosome pairing as a means of estimating the substages of pachytene has been treated in Section III. For a complete account in mammals (and birds), the reader is referred to chapter 3 in this volume. The concern here is that the sex chromosomes are the only ones for which the effects of an absence of meiotic pairing can be studied for their own value. In autosomal situations, as we have seen extensively in the sections dealing with the various chromosome mutations, unpaired chromosomes or segments almost always aggregate with the sex chromosome axial elements. Sex chromosome univalence at pachytene is generally taken to be cell lethal at the first meiotic division and cells seem to succumb at metaphase I (see Reference 10 for additional citations). In a situation where the extent of synapsis between the sex chromosomes was decreased (a finding which accompanies sex chromosome univalence of part of the primary spermatocytes), increased frequencies of severely malformed spermatozoa were found.[88] Moreover, when pairing is more severely affected, as in XOSxr mice that carry the male-determining sequences of a Y chromosomal fragment in the distal telomeric region of the X chromosome, increased frequencies of abnormal spermatozoa (besides a majority of diploid spermatozoa) are observed.[95] Abnormal pairing is achieved here by nonhomologous meiotic pairing between the proximal and distal segments of the X^{Sxr} chromosome: ring formation.

4. The Significance of Associations (Without SC Formation) Between Unpaired Autosomal and Sex Chromosomal Axial Elements for the Impairment of Spermatogenesis

After the discovery that this kind of association could persist into the late diplotene-metaphase I stages of male meiosis in "male-sterile" chromosome mutants (where it was first observed[96]), Forejt placed a heavy weight on this for explaining "sterility" in the male, as contrasted to "fertility" in the female. According to his hypothesis (also see Reference 97), reactivation of the normally transcriptionally inactive X chromosome during male meiosis leads to its partial breakdown and the largely abnormal spermiogenesis. The fact that mouse karyotypes with two X chromosomes (besides Y or Sxr male-determining sequences) do not show any spermatogonial multiplication in the adult seems to reinforce Forejt's explanation. Subsequent research into the levels of the X-linked enzymes PGK-1, G6PD, and HGPRT, which normally are very low in pachytene spermatocytes, but which were clearly increased in a few male sterile karyotypes, including $Ts(5^{12})31H$, lends further support.[98] However, Erasmus and Grootegoed,[109] using T70H/T1Wa sterile males (see Section VIII), observed that the Sertoli cell fraction of pachytene isolates (using Staput and percoll purification) was clearly larger in the mutant than in controls (10% vs. 3%). Since Sertoli cells are known to be metabolically active, this could have disturbed the enzyme readings of the male steriles. Speed[99] expected to obtain direct evidence of sex chromosome activity by measuring ^{3}H uridine incorporation in the sex vesicles of $Ts(1^{13})70H$ and $Ts(5^{12})31H$. Grain counts of "sex vesicles", which, in the second phase pachytene "expanded sex vesicle" stage that was primarily sampled, almost always contain the extra chromosomes (see Section IX), were clearly higher in both mutant karyotypes than in controls. The increase in the oligospermic $Ts(1^{13})70H$ males, however, was greater than in the virtually azoospermic $Ts(5^{12})31H$ males and the grains often appeared to be localized over what most likely was the autosomal element. *In situ* hybridization experiments with X-linked cDNA probes on messenger RNA must give the answer here.

Forejt et al.[36] observed the silver-positive body, which normally resides adjacent to the sex vesicle during late pachytene, to be out of position in the T31H, T43H, and T145H "male sterile" reciprocal translocations. Since the function of this body, although related to rDNA activity, is not precisely known, this interesting observation is difficult to interpret.

It is tempting to associate the general occurrence of malformed spermatozoa and the more incidental high frequencies of diploid spermatozoa[13,95] with Y chromosome function during pachytene, particularly since unpaired autosomal chromosome segments probably remain transcriptionally active despite their sex chromosomal allocyclic chromatin structure. As long as it is not possible to trace the route through which the Y chromosomes influences spermiogenesis, this postulate will be difficult to test. It is of interest to state once more that an absence or reduction of pairing between the sex chromosomes also affects spermiogenesis (Section X.C.3).

5. Why Do X-Autosome Translocations Affect Spermatogenesis Relatively Early?

If allowance is made for the fact that the differentiated areas of the sex chromosomes do not pair stably throughout pachytene, not much difference can be seen at first sight in the meiotic pairing behavior of X-A and A-A translocations. Moreover, X, Y univalence apparently is not a contributive factor since sex chromosome pairing is amazingly normal, even if only a short X-chromosomal segment is available, as in T1R1, T5R1, and T6R1 (Figure 2A). It is even hard to envisage why T2R1, T3R1, and T16H are male-sterile translocations at all because autosomal meiotic pairing is fulfilled. The observation of Searle et al.[47] that there is a correlation between breakpoint position and the stages of germ cell death, could shed some light on male sterile X-A translocations. For this, it should also be remembered that the sex chromosomes exhibit meiotic pairing initiation later than the autosomes. According to Dietrich and de Boer,[24] this delay could be as much as 1 to 2 d (see Figure 1). If some of this delay is also expressed by autosomal segments adjacent to the breakpoints, the pairing difficulties encountered by short interstitial (but also by short translocated) segments would be reinforced. In addition, the autosomal, sex chromosomal chromatin interactions, that in X-A translocations have an extra dimension and always lead to an increased volume of the sex vesicle,[100] could contribute here. The discussion moves much the same way as with the autosomal male sterile conditions (see Section X.C.4). These interactions, which are more intense and, therefore, probably more important, are present in every meiocyte at all pachytene (sub)stages. Moreover, when an X-A translocation allows some degree of spermatogenesis, as with Is1Ct, malformed spermatozoa can be abundantly encountered.[50] More likely, these originate from pachytene meiocytes showing pairing adjustment within the 7;7$^-$ bivalent, leaving the sex vesicle enlarged with the 7-inserted segment. Astonishingly enough, when chr7 pairing is facilitated by the exchange of chr 7$^-$ by a normal 7 (these males are generally known as Is1Ct type II; the former, as type I), spermatogenesis is much improved, sperm counts amounting to 71% of normal. Sperm morphology also shows the usual change to normal.[50] Again, this argues for the ease and completeness of meiotic pairing as a prerequisite for male germ cell survival.

Another exceptional X-autosome rearrangement is the Rb(X;2)2Ad Robertsonian translocation that shows a 74.4% reduction in testicular weight and, as expected by a reduction of this size, is "fertile".[111] Yet unpaired proximal segments of chr 2 are more the rule than the exception.[54] Moreover, features of male sterility are exhibited: the proximal axial element of chr 2 often thickens and some of chr 2 chromatin frequently is included in the sex vesicle. Meiotic pairing initiation, however, is considerably more easy in a Robertsonian trivalent, compared with a reciprocal quadrivalent. It would be interesting to study the kinetics of meiosis in both male-fertile X-autosome rearrangements.

6. Cellular Dependence Within a Symplast

One further difference between male and female mammalian meiosis is that, in male meiosis cells move in synchronous cohorts, the premeiotic, meiotic, and postmeiotic stages being connected by cellular bridges.[101] De Boer and Speed[70] have analyzed the consequences of this phenomenon by virtue of the fact that through a variation in meiotic pairing, male T70H translocation trisomic mice produce both "male-fertile" and "male-sterile" spermatocytes. The latter category, estimated to be $\approx$ 12.5% of the pachytene spermatocytes, occurs randomly over the various symplasts and is "saved" by healthy neighbors. In T31H/ +, a male-sterile translocation with a relatively high frequency of RIV pachytene spermatocytes (see Table 3), these probably are too low in number to boost spermatogenic efficiency. In this respect an old observation of Oakberg and DiMinno[102] might be of significance: radiation-induced cell death during pachytene tended to occur in rows of cells rather than as the random pattern that would be expected. So the effects of meiotic pairing difficulties could be enlarged by the interaction of relatively normal and abnormal cells, the normals not being sufficiently frequent to at least maintain their numbers.

XI. CONCLUSIONS

Despite the relatively fragmentary nature of many chromosome-pairing studies at pachytene in the mouse, the evidence is overwhelming that incompleteness of meiotic pairing is detrimental to meiocyte development in both the male and female. However, the effects are expressed differently, according to the intrinsic properties of gametogenesis in both sexes. In the male, the pachytene stage lasts about twice as long as in the female, the second half postulated to have a function for spermiogenesis. Almost invariably, death of pachytene spermatocytes in male-sterile chromosome mutants commences during this phase. The numbers of diakinesis spermatocytes with contracted bivalents can be conspicuously increased, as are clearly degenerative metaphase Is. Frequencies of malformed spermatozoa vary from moderate ($\approx$ 15%) to very high ($\approx$ 75%), but are usually increased. Alternatively, the second meiotic division may be omitted, which leads to diploid spermatids and spermatozoa that can also be grossly misshapen.[95] It is tempting to blame the generally abnormal spermiogenesis on interaction during the second half of pachytene between sex chromosomal and autosomal chromatin. However, a shortage of pairing between the sex chromosomes can have the same effect. Another factor which probably increases the consequences of meiotic pairing problems in the male, is the fact that pachytene spermatocytes with various degrees of pairing difficulties are found within the same symplast, connected by cellular bridges. If relatively severely affected cells are too numerous, the more nearly normal cells might become exhausted by the process of metabolic cooperation. In the female, after the "death phase" that takes place more synchronously from pachytene to $\approx$ 4 d thereafter, individual meiocytes become surrounded by follicle cells and no extra atresia is subsequently observed (at least in the XO karyotype[17]).

Crossovers seem to be helpful in establishing the paired status of chromosome segments, especially in situations of partner exchange. There are indications that nonhomologous pairing, once achieved during the first half of pachytene, protects the integrity of the meiocyte and its subsequent development.

Some areas open to cytological observations deserve further study. In particular, the sequence of events around zygotene and early pachytene of the various constructs, including X-autosome translocations, could provide better insight into the cell physiological importance of meiotic pairing. Furthermore, it is felt that a selective molecular approach, especially using *in situ* identification of signals (cDNA-mRNA) derived from autosomal- and sex chromosomal loci suffering from pairing problems, is the method of choice for obtaining better insight into the significance of many of the cellular observations set out in this chapter.

ACKNOWLEDGMENTS

Aafke van der Kooi and Elly van Liempt are gratefully acknowledged for preparing the manuscript. We thank Ilse-Dore Adler for her information on Rb(X.2)2Ad and Boudewijn Erasmus and Anton Grootegoed for the use of their preliminary data.

REFERENCES

1. **Counce, S. J. and Meyer, G. F.,** Differentiation of the synaptonemal complex and the kinetochore in *Locusta* spermatocytes studied by whole mount electron microscopy, *Chromosoma,* 44, 231, 1973.
2. **Moses, M. J.,** Synaptonemal complex karyotyping in spermatocyes of the Chinese hamster *(Cricetus griseus).* I. Morphology of the autosomal complement in spread preparations, *Chromosoma,* 60, 99, 1977.
3. **Moses, M. J., Russell, L. B., and Cacheiro, N. L. A.,** Mouse chromosome translocations: visualization and analysis by electron microscopy of the synaptonemal complex, *Science,* 196, 892, 1977.

4. **Fletcher, J. M.,** Light microscope analysis of meiotic prophase chromosomes by silver staining, *Chromosoma,* 72, 241, 1979.
5. **Bloom, S. E. and Goodpasture, C.,** An improved technique for selective silver staining of nucleolar chromosomes, *Hum. Genet.,* 34, 199, 1976.
6. **Speed, R. M.,** Meiosis in the foetal mouse ovary. I. An analysis at the light microscope level using surface-spreading, *Chromosoma,* 85, 427, 1982.
7. **Dresser, M. E. and Moses, M. J.,** Silver staining of synaptonemal complexes in surface spreads for light and electron microscopy, *Exp. Cell Res.,* 121, 416, 1979.
8. **Choi, A. H. C.,** Three dimensional reconstruction of quadrivalents and mapping of translocation breakpoints of the mouse translocations T(2;8)26H and T(9;17)138Ca, *Can. J. Genet. Cytol.,* 22, 261, 1980.
9. **Glamann, J.,** Crossing over in the male mouse as analyzed by recombination nodules and bars, *Carlsberg Res. Commun.,* 51, 143, 1986.
10. **De Boer, P.,** Chromosomal causes for fertility reduction in mammals, in *Chemical Mutagens,* Vol. 10, de Serres, F. J., Ed., Plenum Press, New York, 1986, 427.
11. **De Boer, P., van der Hoeven, F. A., and Chardon, J. A. P.,** The production, morphology, karyotypes and transport of spermatozoa from tertiary trisomic mice and the consequences for egg fertilization, *J. Reprod. Fertil.,* 48, 249, 1976.
12. **Searle, A. G. and Beechey, C. V.,** Sperm count, egg-fertilization and dominant lethality after X-irradiation of mice, *Mutant Res.,* 22, 63, 1974.
13. **De Boer, P., Searle, A. G., van der Hoeven, F. A., de Rooij, D. G., and Beechey, C. V.,** Male pachytene pairing in single and double translocation heterozygotes and spermatogenic impairment in the mouse, *Chromosoma,* 93, 326, 1986.
14. **Searle, A. G., Beechey, C. V., and Evans, E. P.,** Meiotic effects in chromosomally derived male sterility of mice, *Ann. Biol. Anim. Biochim. Biophys.,* 18(2B), 391, 1978.
15. **Searle, A. G., Beechey, C. V., de Boer, P., de Rooij, D. G., Evans, E. P., and Kirk, M.,** A male-sterile insertion in the mouse, *Cytogenet. Cell Genet.,* 36, 617, 1983.
16. **Beechey, C. V., de Boer, P., and van der Hoeven, F. A.,** Egg counts in the ovaries of normal and tertiary trisomic (Ts(1^{13})70H) female mice, *Mouse News Lett.,* 54, 53, 1976.
17. **Burgoyne, P. S. and Baker, T. G.,** Perinatal oocyte loss in XO mice and its implications for the aetiology of gonadal dysgenesis in XO women, *J. Reprod. Fertil.,* 75, 633, 1985.
18. **Oakberg, E. F.,** A description of spermatogenesis in the mouse and its use in analysis of the cycle of the seminiferous epithelium and germ cell renewal, *Am. J. Anat.,* 99, 391, 1956.
19. **Oakberg, E. F.,** Duration of spermatogenesis in the mouse and timing of stages of the cycle of the seminiferous epithelium, *Am. J. Anat.,* 99, 507, 1956.
20. **Monesi, V.,** Autoradiographic study of DNA synthesis and the cell cycle in spermatogonia and spermatocytes of mouse testes, using tritiated thymidine, *J. Cell Biol.,* 14, 1, 1962.
21. **Solari, A. J.,** The spatial relationship of the X and Y chromosomes during meiotic prophase in mouse spermatocytes, *Chromosoma,* 29, 217, 1970.
22. **Moses, M. J.,** New cytogenetic studies on mammalian meiosis, in *Animal Models in Human Reproduction,* Serio, M. and Martini, L., Eds., Raven Press, New York, 1980, 169.
23. **Moses, M. J., Dresser, M. E., and Poorman, P. A.,** Composition and role of the synaptonemal complex, in *Controlling Events in Meiosis,* Evans, C. W. and Dickinson, H. G., Eds., *Symp. Soc. Exp. Biol.,* Vol. 38, Company of Biologists, Cambridge, 1984, 245.
24. **Dietrich, A. J. J. and de Boer, P.,** A sequential analysis of the development of the synaptonemal complex in spermatocytes of the mouse by electron microscopy using hydroxyurea and agar filtration, *Genetica,* 61, 119, 1983.
25. **Collins, A. R. S., Downes, C. S., and Johnson, R. T.,** Introduction: an integrated view of inhibited repair, in *DNA Repair and Its Inhibition,* Collins, A., Downes, C. S., and Johnson, R. T., Eds., IRL Press, Oxford, 1984, 1.
26. **Tres, L. L.,** Extensive pairing of the XY bivalent in mouse spermatocytes as visualized by whole-mount electron microscopy, *J. Cell Sci.,* 25, 1, 1977.
27. **Moses, M. J. and Poorman, P. A.,** Synapsis, synaptic adjustment and DNA synthesis in mouse oocytes, in *Chromosomes Today,* Vol. 8, Bennett, M. D., Gropp, A., and Wolf, U., Eds., Allen and Unwin Ltd., London, 1984, 90.
28. **Oud, J. L., de Jong, J. H., and de Rooij, D. G.,** A sequential analysis of meiosis in the male mouse using a restricted spermatocyte population obtained by a Hydroxyurea/Triaziquone treatment, *Chromosoma,* 71, 237, 1979.
29. **Guitart, M., Coll, M. D., Ponsa, M., and Egozcue, J.,** Sequential study of synaptonemal complexes in mouse spermatocytes by light and electron microscopy, *Genetica,* 67, 21, 1985.
30. **Dietrich, A. J. J. and Mulder, R. J. P.,** A light and electron microscopic analysis of meiotic prophase in female mice, *Chromosoma,* 88, 377, 1983.

31. **Bakken, A. H. and McClanahan, M.,** Patterns of RNA synthesis in early meiotic prophase oocytes from fetal mouse ovaries, *Chromosoma,* 67, 21, 1978.
32. **De Boer, P. and Searle, A. G.,** Summary and synthesis. Workshop on chromosomal aspects of male sterility in mammals, *J. Reprod. Fertil.,* 60, 257, 1980.
33. **Searle, A. G.,** The genetics of sterility in the mouse, in *Genetic Control of Gamete Production and Function,* Crosignani, P. G., Rubin, B. L., and Fraccaro, M., Eds., Academic Press, New York, 1982, 93.
34. **Green, M.,** *Genetic Variants and Strains of the Laboratory Mouse,* Gustav Fischer, Stuttgart, 1981.
35. **Ashley, T., Russell, L. B., and Cacheiro, N. L. A.,** Synaptonemal complex analysis of X-7 translocations in male mice. I. R3 and R5 quadrivalents, *Chromosoma,* 87, 149, 1982.
36. **Forejt, J., Gregorová, S., and Goetz, P.,** XY pair associates with the synaptonemal complex of autosomal male-sterile translocations in pachytene spermatocytes of the mouse, *(Mus musculus), Chromosoma,* 82, 41, 1981.
37. **Beechey, C. V., Kirk, M., and Searle, A. G.,** A reciprocal translocation induced in an oocyte and affecting fertility in male mice, *Cytogenet. Cell Genet.,* 27, 129, 1980.
38. **Burgoyne, P. S., Mahadevaiah, S., and Mittwoch, U.,** A reciprocal autosomal translocation which causes male sterility in the mouse also impairs oogenesis, *J. Reprod. Fertil.,* 75, 647, 1985.
39. **Mittwoch, U., Mahadevaiah, S., and Olive, M. B.,** Retardation of ovarian growth in male-sterile mice carrying an autosomal translocation, *J. Med. Genet.,* 18, 414, 1981.
40. **Mahadevaiah, S., Mittwoch, U., and Moses, M. J.,** Pachytene chromosomes in male and female mice heterozygous for the Is(7;1)40H insertion, *Chromosoma,* 90, 163, 1984.
41. **Mittwoch, U., Mahadevaiah, S., and Setterfield, L. A.,** Chromosomal anomalies that cause male sterility in the mouse also reduce ovary size, *Genet. Res.,* 44, 219, 1984.
42. **De Boer, P., Stam, P., van Oosten, L., and Wauben-Penris, P. J. J.,** Chiasma frequency and position in male and female mice of chromosomes involved in homozygous and heterozygous translocations and tertiary trisomy, *Genetica,* 61, 205, 1983.
43. **Ashley, T., Russell, L. B., and Cacheiro, N. L. A.,** Synaptonemal complex analysis of X-7 translocations in male mice: R2 and R6 quadrivalents, *Chromosoma,* 88, 171, 1983.
44. **Ashley, T.,** Nonhomologous synapsis of the sex chromosomes in the heteromorphic bivalents of two X-7 translocations in male mice: R5 and R6, *Chromosoma,* 88, 178, 1983.
45. **Ashley, T.,** Application of the spreading techniques to structural heterozygotes, in *Chromosomes Today,* Vol. 8, Bennett, M. D., Gropp, A., and Wolf, U., Eds., George Allen and Unwin Ltd., London, 1984, 80.
46. **Ashley, T. and Russell, L. B.,** A new type of nonhomologous synapsis in T(X;4)1Rl translocation male mice, *Cytogenet. Cell Genet.,* 43, 194, 1986.
47. **Searle, A. G., Beechey, C. V., Evans, E. P., and Kirk, M.,** Two new X-autosome translocations in the mouse, *Cytogenet. Cell Genet.,* 35, 279, 1983.
48. **Chandley, A. C.,** A model for effective pairing and recombination at meiosis based on early replicating sites (R-bands) along chromosomes, *Hum. Genet.,* 72, 50, 1972.
49. **Eicher, E. M.,** X-autosome translocations in the mouse: total inactivation versus partial inactivation of the X-chromosome, *Adv. Genet.,* 15, 175, 1970.
50. **Meistrich, M. L., Göhde, W., White, R. A., and Longtin, J.,** ''Cytogenetic'' studies of spermatids of mice carrying Cattanach's translocation by flow cytometry, *Chromosoma,* 74, 141, 1979.
51. **Moses, M. J., Karatsis, P. A., and Hamilton, A. E.,** Synaptonemal complex analysis of heteromorphic trivalents in Lemur hybrids, *Chromosoma,* 70, 141, 1979.
52. **Brown, B. B. and Burtenshaw, M. D.,** Pachytene pairing in Robertsonian heterozygotes, *Mouse News Lett.,* 62, 70, 1980.
53. **Winking, H. and Johannisson, R.,** Pattern of pachytene pairing in mouse hybrids with chain and ring multivalents, *Clin. Genet.,* 17, 94, 1980.
54. **Adler, I.-D., Schmöller, R., Nether, B., and Johannisson, R.,** A Robertsonian translocation involving the X-chromosome of the mouse, *Genet. Res.,* 49, 249, 1987.
55. **Cattanach, B. M. and Moseley, H.,** Non-disjunction and reduced fertility caused by the tobacco mouse metacentric chromosomes, *Cytogenet. Cell Genet.,* 12, 264, 1973.
56. **Poorman, P. A., Moses, M. J., Davisson, M. T., and Roderick, T. H.,** Synaptonemal complex analysis of mouse chromosomal rearrangements. III. Cytogenetic observations on two paracentric inversions, *Chromosoma,* 83, 419, 1981.
57. **Tease, C. and Fisher, G.,** Further examination of the production-line hypothesis in mouse foetal oocytes. I. Inversion heterozygotes, *Chromosoma,* 93, 447, 1986.
58. **Davisson, M. T., Poorman, P. A., Roderick, T. H., and Moses, M. J.,** A pericentric inversion in the mouse, *Cytogenet. Cell Genet.,* 30, 70, 1981.

59. **Gropp, A. and Winking, H.,** Robertsonian translocations: cytology, meiosis, segregation patterns and biological consequences of heterozygosity, in *Biology of the House Mouse,* Berry, R. J., Ed., Academic Press, London, 1981, 141.
60. **Moses, M. J. and Poorman, P. A.,** Synaptonemal complex analysis of mouse chromosomal rearrangements. II. Synaptic adjustment in a tandem duplication, *Chromosoma,* 81, 519, 1981.
61. **Wauben-Penris, P. J. J., van der Hoeven, F. A., and de Boer, P.,** Chiasma frequency and non-disjunction in heteromorphic bivalents: meiotic behaviour in T(1;13)70H/T(1;13)Wa mice as compared to T(1;13)70H/ T(1;13)70H mice, *Cytogenet. Cell Genet.,* 36, 547, 1983.
62. **Gropp, A., Winking, H., and Redi, C.,** Consequences of Robertsonian heterozygosity segregation impairment of fertility versus male-limited sterility, in *Genetic Control of Gamete Production and Function,* Crosignani, P. G., Rubin, B. L., and Fraccaro, M., Eds., Academic Press, New York, 1982, 115.
63. **Redi, C. A., Garagna, S., Hilscher, B., and Winking, H.,** The effects of some Robertsonian chromosome combinations on the seminiferous epithelium of the mouse, *J. Embryol. Exp. Morphol.,* 85, 1, 1985.
64. **Wolf, K. W. and Winking, H.,** Topography of pachytene nuclei from male NMRI and complex heterozygous mice *(Mus musculus),* personal communication.
65. **Chandley, A. C.,** A pachytene analysis of two male-fertile paracentric inversions in chromosome 1 of the mouse and in the male-sterile double heterozygote, *Chromosoma,* 85, 127, 1982.
66. **De Boer, P. and Branje, H. E. B.,** Association of the extra chromosome of tertiary trisomic male mice with the sex chromosomes during first meiotic prophase, and its significance for impairment of spermatogenesis, *Chromosoma,* 73, 369, 1979.
67. **Mahadevaiah, S. and Mittwoch, U.,** Synaptonemal complex analysis in spermatocytes and oocytes of tertiary trisomic Ts(5^{12})31H mice with male sterility, *Cytogenet. Cell Genet.,* 41, 169, 1986.
68. **Setterfield, L. A. and Mittwoch, U.,** Reduced oocyte numbers in tertiary trisomic mice with male sterility, *Cytogenet. Cell Genet.,* 41, 177, 1986.
69. **De Boer, P. and van Beek, M. E. A. B.,** Meiosis of T70H translocation trisomic male mice. I. Meiotic configurations and segregation, *Chromosoma,* 87, 303, 1982.
70. **De Boer, P. and Speed, R. M.,** Meiosis of T70H translocation trisomic male mice. II. Meiotic rate, spermatocyte interactions and fertility, *Chromosoma,* 87, 315, 1982.
71. **De Boer, P. and van der Hoeven, F. A.,** The use of translocation-derived "marker-bivalents" for studying the origin of meiotic instability in female mice, *Cytogenet. Cell Genet.,* 26, 49, 1980.
72. **Forejt, J.,** X-Y involvement in male sterility caused by autosome translocations— a hypothesis, in *Genetic Control of Gamete Production and Function,* Crosignani, P. G., Rubin, B. L., and Fraccaro, M., Eds., Academic Press, New York, 1982, 135.
73. **Speed, R. M. and de Boer, P.,** Delayed meiotic development and correlated death of spermatocytes in male mice with chromosome abnormalities, *Cytogenet. Cell Genet.,* 35, 257, 1983.
74. **Stern, H. and Hotta, Y.,** Molecular biology of meiosis: synapsis-associated phenomena, in, *Aneuploidy, Etiology and Mechanisms,* Dellarco, V. L., Voytek, P. E., and Hollaender, A., Eds., Plenum Press, New York, 1985, 305.
75. **De Boer, P. and Nijhoff, J. H.,** Meiosis of male Rb(11.13)4Bnr/ + , Ts(1^{13})70H mice, *Mouse News Lett.,* 70, 105, 1984.
76. **Wauben-Penris, P. J. J., van Meel, G. P. J. M., and de Boer, P.,** Spermatogenic delay and increased chiasma frequency in T70H/ + male mice with hydroxyurea-Trenimon limited spermatocyte populations, *Can. J. Genet. Cytol.,* 27, 192, 1985.
77. **Beechey, C. V. and Speed, R. M.,** Double tertiary trisomy in T(5;12)31H, *Mouse News Lett.,* 64, 56, 1981.
78. **Winking, H. and Johannisson, R.,** Meiosis and testis histology of hybrids of the genus *Mus, Hereditas,* 104, 169, 1986.
79. **De Boer, P.,** Male meiotic behaviour and male and female litter size in mice with the T(2;8)26H and T(1;13)70H reciprocal translocations, *Genet. Res.,* 27, 369, 1976.
80. **Gregorová, S., Baranov, V. S., and Forejt, J.,** Partial trisomy (including T-t gene complex) of the chromosome 17 of the mouse. The effect on male fertility and the transmission to progeny, *Folia Biol. Prague,* 27, 171, 1981.
81. **Rasmussen, S. W.,** Chromosome pairing in triploid females of *Bombyx mori* analyzed by three dimensional reconstructions of synaptonemal complexes, *Carlsberg Res. Commun.,* 42, 163, 1977.
82. **Monesi, V.,** Synthetic activities during spermatogenesis in the mouse, *Exp. Cell Res.,* 39, 197, 1965.
83. **Kierszenbaum, A. L. and Tres, L. L.,** Nucleolar and perichromosomal RNA synthesis during meiotic prophase in the mouse testis, *J. Cell Biol.,* 60, 39, 1974.
84. **Oud, J. L. and Reutlinger, A. H. H.,** The behaviour of silver-positive structures during meiotic prophase of male mice, *Chromosoma,* 81, 569, 1981.
85. **Russell, L. D. and Clermont, Y.,** Degeneration of germ cells in normal, hypophysectomized and hormone treated hypophysectomized rats, *Anat. Rec.,* 187, 347, 1977.

86. **Willison, K. R. and Ashworth, A.,** Mammalian spermatogenic gene expression, *Trends Genet.*, 3, 351, 1987.

87. **Krzanowska, H.,** Inheritance of sperm head abnormality types in mice — the role of the Y chromosome, *Genet. Res.*, 28, 189, 1976.

88. **De Boer, P. and Nijhoff, J. H.,** Incomplete sex chromosome pairing in oligospermic male hybrids of *Mus musculus* and *M. musculus molossinus* in relation to the source of the Y chromosome and the presence or absence of a reciprocal translocation, *J. Reprod. Fertil.*,62, 235, 1981.

89. **Lyon, M. F.,** Mouse chromosome atlas, *Mouse News Lett.*, 78, 12, 1987.

90. **Krzanowska, H.,** Interstrain competition amongst mouse spermatozoa inseminated in various proportions, as affected by the genotype of the Y chromosome, *J. Reprod. Fertil.*, 77, 265, 1986.

91. **Bishop, C. E. and Hatat, D.,** Molecular cloning and sequence analysis of a mouse Y chromosome RNA transcript expressed in the testis, *Nucleic Acids Res.*, 15, 2959, 1987.

92. **Somssich, I., Hameister, H., and Winking, H.,** The pattern of early replicating bands in the chromosomes of the mouse, *Cytogenet. Cell Genet.*, 30, 222, 1981.

93. **Richler, C., Uliel, E., Kerem, B.-S., and Wahrman, J.,** Regions of active chromatin conformation in "inactive" male meiotic sex chromosomes of the mouse, *Chromosoma*, 95, 167, 1987.

94. **Rajcan Separovic, E. and Chandley, A. C.,** Lack of evidence that the X_qY_q pairing tips at meiosis in the mouse show hypersensitivity to DNAse I, *Chromosoma*, 95, 290, 1987.

95. **Levy, E. R. and Burgoyne, P. S.,** Diploid spermatids: a manifestation of spermatogenic impairment in XO Sxr and T31H/ + male mice, *Cytogenet. Cell Genet.*, 42, 159, 1986.

96. **Forejt, J.,** Non-random association between a specific autosome and the X chromosome in meiosis of the male mouse: possible consequence of homologous centromeres' separation, *Cytogenet. Cell Genet.*, 13, 369, 1974.

97. **Lifschytz, E.,** X-chromosome inactivation: an essential feature of normal spermiogenesis in male hetero-gametic organisms, in *The Genetics of the Spermatozoon*, Beatty, R. A. and Gluecksohn-Waelsch, S., Eds., Beatty and Gluecksohn-Waelsch, Edinburgh, 1972, 223.

98. **Hotta, Y. and Chandley, A. C.,** Activities of X-linked enzymes in spermatocytes of mice rendered sterile by chromosomal alterations, *Gamete Res.*, 6, 65, 1982.

99. **Speed, R. M.,** Abnormal RNA synthesis in sex vesicles of tertiary trisomic male mice, *Chromosoma*, 93, 267, 1986.

100. **Solari, A. J.,** The behaviour of chromosomal axes in Searle's X-autosome translocation, *Chromosoma*, 34, 99, 1971.

101. **Fawcett, D. W.,** Observations on cell differentiation and organelle continuity in spermatogenesis, in *The Genetics of the Spermatozoan*, Beatty, R. A. and Gluecksohn-Waelsch, S., Eds., Beatty and Gluecksohn-Waelsch, Edinburgh, 1972, 37.

102. **Oakberg, E. F. and DiMinno, R. L.,** X-ray sensitivity of primary spermatocytes of the mouse, *Int. J. Radiat. Biol.*, 2, 196, 1960.

103. **Forejt, J. and Gregorová, S.,** Meiotic studies of translocations causing male sterility in the mouse. I. Autosomal reciprocal translocations, *Cytogenet. Cell Genet.*, 19, 159, 1977.

104. **De Boer, P. and Groen, A.,** Fertility and meiotic behaviour of male T70H tertiary trisomics of the mouse *(Mus musculus)*. A case of preferential telomeric meiotic pairing in a mammal, *Cytogenet. Cell Genet.*, 13, 489, 1974.

105. **Zwanenburg, T. S. B., de Boer, P., and Stam, P.,** Clonal analysis of radiation induced translocation in stem-cell spermatogonia of normal and T70H translocation heterozygous mice, *Mutat. Res.*, 83, 207, 1981.

106. **De Rooij, D. G.,** personal communication.

107. *Mouse News Lett.*, 77, 1987.

108. **De Boer, P.,** unpublished.

109. **Erasmus, B. and Grootegoed, A.,** personal communication.

110. **DeRooij, D. G.,** personal communication.

111. **Adler, I.-D.,** personal communication.

Chapter 3

SEX CHROMOSOME PAIRING AND FERTILITY IN THE HETEROGAMETIC SEX OF MAMMALS AND BIRDS

Alberto J. Solari

TABLE OF CONTENTS

I. INTRODUCTION: THE DISCOVERY OF PARTIAL SYNAPSIS OF THE SEX CHROMOSOMES IN EUTHERIAN MAMMALS

In 1934, Koller and Darlington[1] described a mechanism of sex-chromosome association and disjunction during meiosis in *Rattus norvegicus* on the basis of observations on sectioned tissues fixed in Bouin-Allen mixture. The authors assumed that the X and Y chromosomes were acrocentric, with sizable short arms, and that during pachytene, both chromosomes were paired through the short arms, the kinetochores, and a proximal piece of the long arms, while the remaining parts were considered as "differential segments" which were not able to pair. According to this view, chiasmata might occur either in the short arms or beyond the kinetochores, the first ones giving "prereduction" (90%) and the latter giving "post-reduction" (10%) of the sex chromosomes in anaphase I. In a series of papers (see Mittwoch[2]), Koller extended this view to a number of mammals — including man — and proposed that this behavior was the general one of mammalian X and Y chromosomes.

However, Koller's view, which was founded on the hypothesis that chiasmata are needed for regular segregation of chromosomes,[2] was based on ambigous cytological data that were not confirmed by other authors. Thus, Matthey[3] remarked that "postreduction" of sex chromosomes was not observed in rat meiosis, that pairing of the X and Y chromosomes could not be proved, and that even the diploid numbers of some of the species described by Koller had not been confirmed. The later work of some cytologists, notably Sachs[4,5] and Ohno et al.,[6] described the special behavior of the X and Y chromosomes during pachytene in rodents and in man. In these species — as in most mammals — the X and Y chromosomes form a heteropycnotic body, the so-called "sex vesicle", better described as the XY body.[7] This condensed, intranuclear body could not clearly show if pairing of the X and Y chromosomes existed. Furthermore, cells prepared by squashing or spreading clearly showed that the association between the X and Y at metaphase I was "end-to-end" in most mammalian species, giving support to the assumption that this association was nonchiasmatic. Thus, up to the late 1960s the predominant hypothesis was that the X and Y chromosomes did not pair — at least not in the same way as autosomes — and that chiasmata between these chromosomes were exceptional.[2] "Partial sex-linkage" remained an unverified hypothesis.[2]

Following the discovery of the synaptonemal complex (SC) by Moses,[8] and the realization that the SC represents the fine structure of synapsis, a new approach to the problem of sex-chromosome association in male mammalian meiosis was developed[9-11] (reviewed by Solari[7] and by Moses[12]). The enigmatic structure called "sex vesicle", Lenhossek's body, or other names,[7] was first identified in electron micrographs of mouse spermatocytes as an XY pair.[9] This work definitively showed that the "sex vesicle" had no surrounding membranes, and that it had a structure formed by typically packed chromatin fibers and "axes" that were related to the lateral components of SCs.[9] Thus, it was possible to analyze the pairing of sex chromosomes by the study of their fine structure.[9] This analysis required information on the full structure of the chromosomes, and thus serial thin-sectioning and three-dimensional reconstruction techniques were applied to the reconstruction of the XY pair.[13] The first results of these reconstructions showed the existence of an SC in the XY pair of the mouse.[11,13] Working with single (thin and thick) sections, Ford and Woollam[14] confirmed Solari's observations on the mouse and showed that an SC could be seen in the XY body of the golden hamster (however, see the exceptionality of the hamster group in Section V). Results of three-dimensional reconstructions accumulated between 1966 and 1971 (reviewed by Solari[7] and by Westergaard and von Wettstein[15]), proved beyond reasonable doubt that partial synapsis was the regular behavior of the X and Y chromosomes in the most-studied eutherian mammals, including man.[7] The assumption of Koller and Darlington about "pairing" and "differential" segments in the mammalian XY pair, was finally validated by the discovery of the regular formation of SCs between the X and Y axes at pachytene,[7] thus demonstrating the validity of the ultrastructural approach to cytogenetics.

Almost as debated as XY synapsis, the nature of the end-to-end association of the XY pair at metaphase I could not be decided by cytological techniques.[2] The study of the fine structure of autosomal chiasmata in mice[16] and the finding of a similar behavior in the XY body,[7] suggested the chiasmatic nature of this association.[7] The recognition of "recombination nodules" (RNs)[17] and their location in XY pairs[18,19] provided additional support to this view. However, the matter continued to be controversial[20] until the recent application of sex chromosome-specific DNA probes that strongly support the chiasmatic nature of this association (see Section V).

The special condensation of the sex chromosomes during male meiosis has been known for a long time.[7] This condensation is widespread, but not universal, during meiosis in the heterogametic sex.[7] Sachs[5] assumed that this condensation was needed for the restriction of recombination between the X and Y chromosomes. Furthermore, Lifschytz and Lindsley[21] presented the hypothesis that the inactivation of the X chromosome, as shown in the XY condsensed body, was a requisite for the development of spermatogenesis. From these functional considerations, some mechanisms have been assumed to involve sex-chromosome pairing and infertility (see Section VIII).

This review is not intended to be comprehensive; rather, the aim is to stress the lasting continuity of some ideas, as well as some recent developments. Previous reviews on chromosome pairing,[7,22] as well as other chapters of this book, provide an additional background of references.

II. HYPOTHESES ON THE PHYLOGENESIS OF SEX CHROMOSOMES OF MAMMALS AND BIRDS

Differentiated sex chromosomes originate from an ancestral, autosomal pair.[23] During the evolutionary process, segments of the original autosomes may remain relatively unchanged, thus constituting "relics" of the ancestral pair. If sex chromosomes have a pairing region, it seems reasonable to assume that this region is (or contains) the relics of the ancestral pair.[2] However, this is not always true in mammals, as shown by some hamsters and deer mice (see Section V). The evolution of sex chromosomes is assumed to proceed towards the conversion of "homologous" regions into "differential" regions,[1] the latter sharing little or no genetic content. Again, this is far from being proved. The Y member of the sex pair is generally reduced in size, while the X remains constant.[24] In mammals, the X chromosome is extensively conserved, as shown by cytological as well as biochemical criteria.[24,25] In mammals, the "original" (conserved) X chromosome amounts to 5 to 5.6% of the homogametic haploid set,[26] while the "original" Z chromosome of birds amounts to 7.2 to 9.3% of the haploid genome.[27] The reasons for the evolutionary conservation of the X (or Z) chromosome are not well understood;[28] they could be related to the widespread sterility of X-autosome translocations in mammals.[28] However, Z-autosome translocations do not result in gametogenic impairment (see Section IX) and, thus, the negative selection against these rearrangements cannot explain the "conservation" of the Z chromosome.

There is now consistent evidence that the mammalian X and avian Z chromosomes are not directly related, and that they might have originated independently. Thus, the genic contents of the X and the Z chromosomes are very different[29,30] and several of the "conserved" genes of the mammalian X are distributed among the autosomes of the chicken. On the other hand, in both the human and in the chicken, the heteromorphic sex pair (XY or ZW) has the unusual feature of undergoing "obligatory" recombination in a very limited region close to the tips of the chromosomes (see Section V). From a functional viewpoint, while the mammalian XY pair becomes condensed and inactive during meiosis, the ZW pair does not become condensed (see Section IX) and is assumed to remain as active as autosomes. Regarding somatic cells, dosage compensation by X-chromosome inactivation is the property of mammals since no dosage compensation of Z-linked genes has been observed.[29]

There is scant information concerning the evolutionary relationships between the sex chromosomes of eutherian, metatherian, and prototherian mammals. While random (paternal or maternal) X-chromosome inactivation is found in the somatic cells of eutherian females,[31] in marsupials, the paternal X is preferentially inactivated,[32] thereby establishing a basic difference between these groups. Additionally, the meiotic behavior of the XY pair in marsupials is markedly different from that of eutherians (see Section VI). Prototherian mammals seem to have special sex-chromosome mechanisms.[33] A general perspective on the evolution of sex chromosomes has been provided by Bull.[34]

III. THE XY PAIR (XY BODY, "SEX VESICLE")

A. Morphology

The typical XY pair is a heteropycnotic body attached to the nuclear envelope and easily visible during the pachytene stage (reviewed by Solari). The size of the XY body is related to that of the sex chromosomes and may reach a considerable size in species having very large sex chromosomes, such as *Microtus agrestis*[35] or the indian muntjac, *Muntiacus muntjak*.[36] In the human, the XY body has a maximum diameter of 3.5 μm.[10,37] In some species, e.g., the primate *Cebus apella*,[38] the XY body is not especially prominent and can be distinguished mainly by its strong staining affinity. The location of the XY body may be studied in sections. In this way, the close relationship of the XY body and the nuclear envelope is observed.[7] This association is due to the end-attachment of the X and Y axes to the nuclear envelope;[7] thus, when these attachments move at early pachytene, the shape of the XY body is more variable.[13] Unlike nucleoli, the XY body is seldom spherical, but is generally oblong and has an irregular outline. Its staining affinity is variable among different species and during the substages of pachytene. The XY body is stained with Feulgen and basic stains, and histochemical techniques do not show the presence of RNA as a regular constituent.[39] With basic stains, the XY body generally stains with an intensity different from that of other chromatin pieces. In rodents, the XY body generally stains lightly and more homogenously than autosomes in semi-thin sections (0.5 μm thick, plastic embedding). Little structural detail is apparent in the XY body after the conventional techniques of squashing and spreading (see Solari[7]). However, a comparison of light- and electron micrographs shows that the slits seen with the light microscope correspond to the X and Y axes.[7] When silver stains[40,41] are used, the X and Y axes are sharply outlined, and even some fine structural details can be shown.

B. Occurrence

An XY body has been seen in the pachytene spermatocytes of most species among eutherian mammals. A few species lacking an XY body have been reported. These exceptional cases include pathological (desynaptic) spermatocytes,[42] sterile hybrids,[43] particular species or groups in which the X or the Y chromosomes are translocated on autosomes in their natural condition,[44,45] and ectopic germ cells.[46] Thus, the occurrence of the XY body seems to be a regular feature of mammalian male meiosis whenever no rearrangements with autosomes are involved and meiosis proceeds in the normal way. Several of these cases will be discussed in Section VII.

One of the best-known species lacking an XY body is the marsupial *Lagorchestes conspicillatus*.[47] In this species, an autosome (A1) is translocated to the original X chromosome, and the Y is compounded by the other autosomal homolog at one end and by an additional translocation of a second autosome (A2) at the other end. A distinct "sex vesicle" was not observed at pachytene. Instead, a triradiate structure was present in which two arms had autosomal appearance, and a third arm was heteropycnotic and had an associated nucleolus (the bifurcating region, as shown in Figure 5b of the original paper,[47] is also negatively heteropycnotic). In silver-stained pachytene spermatocytes, most of the heteropycnotic arm and a short segment of the other arm had thickened axial regions which represent most or all of the original X and Y chromosomes. Furthermore, the original X and Y chromosomes, as shown by their heteropycnotic appearance, do not show physical attachment at diplotene and metaphase I, but the autosomal pieces show distinct chiasmata. It was assumed[47] that the capacity to form a "sex vesicle" (XY body) was lost as a result of one or more of the three translocations involved in the production of this chromosome system. However, several of the characteristics of the XY body persist, as shown by the thickened axial regions and heteropycnotic appearance. It would be interesting to test the transcriptional activity over the regions having thickened axes, as well as the fine structure of this chromatin. Such information could show whether, despite the presence of the flanking autosomal chromatin, the original X and Y chromatin retained their ability to become inactive at pachytene in spermatogenesis.

Other instances of species having translocated sex chromosomes in which a distinct XY body was not observed include the phyllostomid bats[44] (see Section VII) and the Indian mongoose.[45] However, the aforementioned remarks about *L. conspicillatus* should be applicable. Thus, the assumption, that, even when joined to autosomes, the pieces of sex chromosomes behave in the "right" way might not be discarded. In fact, this happens in male mice heterozygous for Searle's T(X;16)16H translocation.[48] An important difference between the sterile mice carrying Searle's translocation and the fertile species with translocated sex chromosomes might rest on the establishment of a barrier to the spreading of inactivation from the sex chromosomes toward the autosomes in the latter species (see Section VII).

C. Relationships with Nucleoli

The XY body may be associated with a nucleolus or nucleolar-derived-body,[7,49] but this association is not inherent to the nature of the XY body[7] and no association is present in many mammalian species.[7] The best known association between the XY body and a nucleolus is that of the common mouse.[39,50] A regular association between the XY body and a conspicuous, RNA-containing body[39] develops at mid-pachytene. The fine structure of this body shows the presence of segregated, granular, fibrillar, and amorphous regions.[50] The "oblong body" inside this nucleolar body has been identified as the fibrillar center,[51] which is argirophilic[39] and positive for the NOR-silver techniques.[51] In the mouse, NORs are located in three to five autosomal pairs,[52] disproving a previous assumption about the location of NORs in the sex chromosomes. The behavior of autosomal NORs during pachytene has been studied with electron microscopy and cytochemical techniques,[51] showing that several nucleoli develop at zygotene-early pachytene, in association with autosomes. At mid-pachytene, one or more NORs move toward the inner side of the XY body, allegedly without detaching from the autosomes.[51] The nucleolar body grows during late pachytene and forms a cap on the inner side of the XY body.[50] Its segregated structure is similar to that of inactive nucleoli,[51] although ^{3}H-uridine incorporation has been observed at mid-pachytene. There is not yet a clear understanding of the mechanisms involved in the movement of NORs towards the XY body.

D. Metabolism

The XY body shows a very low metabolic activity.[7] Very low levels of ^{3}H-uridine incorporation on the XY body have been observed repeatedly.[53,54] Protein turnover also exists at low levels.[55] An unusual amount of RNA synthesis in the XY bodies of two different tertiary trisomic mice (Ts[1-13]70H and Ts[5-12]31H) has been reported by Speed.[56] In these animals, an extra, small marker chromosome is usually associated with the XY pair at pachytene.[57] These trisomic mice show a degree of reduced fertility that depends on their genetic background.[57] After incubation with ^{3}H-uridine, the average grain count per XY body was significantly raised in all trisomics, compared with controls. This transcriptional activity might have resulted from the marker chromosomes remaining active, although other mechanisms are not excluded.[56]

The recognition of regular recombination in the pairing region, and the presence of a "pseudoautosomal region" (see Section V) has rekindled interest in the metabolism of the XY body. Sensitivity to DNase I is assumed to reveal chromatin regions which are active, or potentially active, in transcription.[58] In human spermatocytes both at pachytene and metaphase I, DNase I detects hypersensitive sites specific for the XY pair.[59] These sites are the pairing region (Xp ter and Yp ter), the telomeric regions of the long arms (Xq-Yq), and a site just below the centromere in Xq.[59] A functional relationship has been suggested among hypersensitivity to DNase I, early replication, and the presence of high recombinational activity, perhaps owing to a necessary opening up of chromatin to allow for the recombination process at meiotic prophase.[59] In this respect, it has been shown that the earliest replicating regions in human mitotic chromosomes are Xp22.13 — Xp22.3 and Yp11.2 — Yp11.32.[60] The early replicating region of human cells is evolutionarily conserved, as shown by its presence in the X chromosome of the great apes.[61] This early replication region is involved in pairing (see Section IV).

E. Mechanism of Formation

Although the presence of the XY body in spermatocytes is almost universal among mammals,[7] this body is not found in other vertebrates having heteromorphic sex chromosomes. Since mammals have a characteristic mechanism of gene dosage compensation by X-chromosome inactivation in the somatic cells of females, it has been suggested that XY body formation and sex chromatin are related processes, probably sharing some underlying mechanism.[7] Since the Y chromosome is generally a small chromosome, the mass of the XY body is formed mainly by the condensed X chromosome. If inactivation of the single X chromosome is needed for the normal development of mammalian spermatogenesis, as assumed by Lifschytz and Lindsley,[21] the sharing of a basic mechanism for spermatogenesis and gene dosage compensation might be advantageous in economizing biological information. As previously pointed out,[7] the peculiar packing of chromatin fibers in the XY body of the mouse spreads out to an autosomal region which is joined to the distal part of the X chromosomal piece in Searle's T(X;16)16H translocation.[48] This effect mimics the "spreading of inactivation" effect found in somatic cells.[7] The molecular mechanisms involved both in sex-chromatin formation and in the formation of the XY body remain largely unknown. Thus, it is difficult to assess the information on the reported difference in transforming ability of X-DNA from spermatogenic cells and from somatic cells.[62]

The time at which X-chromosome inactivation occurs in male meiosis is bracketed between the pre-leptotene spermatocyte and pachytene.[7] In human spermatogonia, there is no evidence of heteropycnosis or condensation of the X chromosome[7] and the X chromosome of the mouse shows only a paracentromeric, late-replicating region.[63] The cycle of the seminiferous epithelium in rodents (see Section IV.B) gives an accurate timetable for studies with thin sections. Using this technique, the characteristic condensation of chromatin fibers of the XY body is not apparent before zygotene in mouse spermatocytes.[50] This observation is repeated in other species.[7] Thus, the available evidence suggests that X-chromosome condensation is concurrent with the development of axes in meiotic chromosomes. This assembly of axial material is not, per se, instrumental in chromatin inactivation, as shown by the very actively transcribing autosomal bivalents.[55] In most mammals, the development of the X and Y axes is immediately followed by pairing (see Section IV). However, an SC is not required for the formation of the XY body, as shown both in some eutherians (e.g., the sand rat *Psammomys obesus*) and all marsupials (see Section VI). Thus, a simple telomere association or the presence of Y chromatin in the same nucleus is able to trigger the condensation of the X chromosome when it has reached a responsive stage. The behavior of ectopic germ cells has recently added some information on this subject. In mouse fetal adrenal glands, some germ cells are present in both sexes before and about the time of birth.[64] These ectopic germ cells enter meiotic prophase at the same time as the ovarian germ cells, regardless of the sex of the individual, and differentiate into oocytes, many of which degenerate at meiotic prophase.[64] These scarce germ cells (1 to 13 per adrenal) suggest that all germ cells are potentially able to differentiate in the female direction.[64] In Giemsa-stained, air-dried preparations, these cells do not show the "sex vesicle" both in male and female fetuses,[46] and this has suggested that the development of the XY body is associated with spermatogenesis as such.[46] However, it is not known which is the pairing state of the sex chromosomes in these cells; it is only assumed from the absence of an XY body. The X and Y chromosomes could be paired without forming a condensed body, or else they could simply be unpaired. In the former, the unusual case of unequal sex chromosomes, paired without condensation, would occur; in the latter, more typical case, the lack of an XY body would simply mean that they do not pair. When a general impairment of pairing exists, it is common to observe the lack of an XY body.[42] This is also suggested by the frequent degeneration that occurs in these ectopic germ cells.[64] Thus, these interesting ectopic cells deserve a thorough investigation.

In X0 Sxr' mice, spermatogenesis is severely affected, but occasional tubules show pachytene spermatocytes in which a clear "sex vesicle" has been observed,[65] indicating that this structure can form in the absence of H-Y antigen.[65] There seems to be little support for the assumption that formation of the XY body depends on a spermatogenic environment, regardless of the chromosome constitution of the cell. Thus, spermatogenesis proceeds in many vertebrates, as in birds, without forming a heteropycnotic body. In fact, oocytes develop a "sex vesicle" in an ovary in some *Akodon* field mice. Thus, it may be concluded that most of the evidence suggests that formation of the XY body depends on an interaction between segments of the X and Y chromosomes, which triggers a built-in mechanism for the condensation of the X chromosome and perhaps activates a similar mechanism in the Y chromosome.

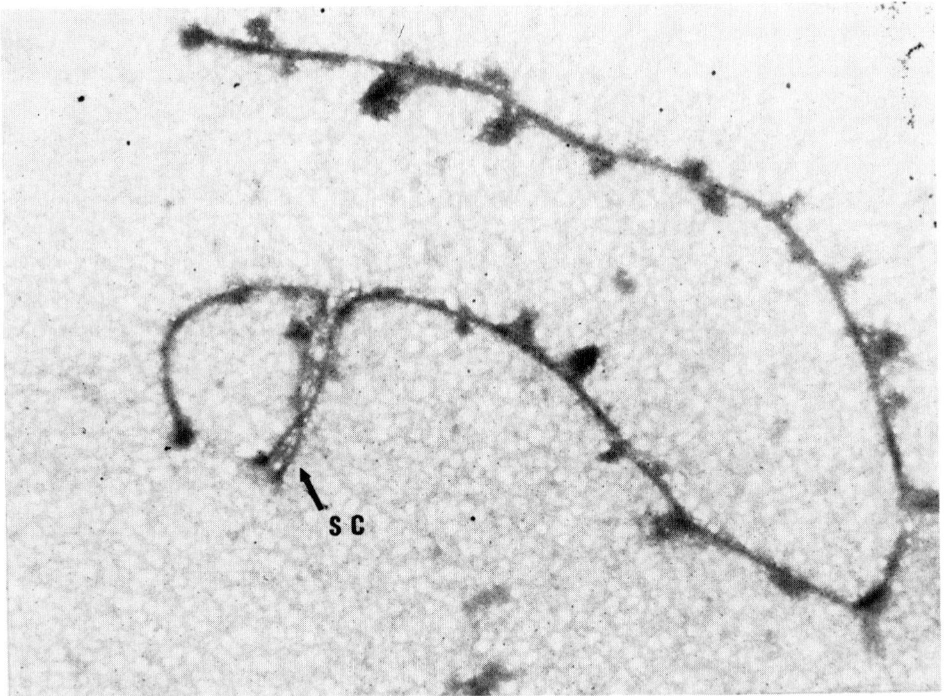

FIGURE 1.   Normal human XY pair. This XY pair has a relatively long synaptonemal complex (SC), which stretches along 38% of the short (Y) axis. A number of typical excrescences are found along the X and Y axes, but neither branching nor extensive splitting is observed in the axes. These features are characteristic of the human XY pair at early pachytene (see discussion in text and in Solari[19]). Some excrescences are found in the nontelomeric half of the SC, suggesting the transient nature of this segment of the SC. Microspread preparation stained with phosphotungstic acid. (Magnification × 12,500.)

IV. PARTIAL SYNAPSIS OF THE X AND Y CHROMOSOMES

A. General Scheme of Axial Behavior

The use of an ultrastructural approach (see Introduction) has been most fruitful for the investigation of the XY pair. Thus, it is presently accepted that chromosomal axes faithfully represent chromosome behavior.[7,22] The behavior of the X and Y axes has been intensively studied in the mouse,[13,67] human,[19,37] Chinese hamster,[68] and rat.[60,70] These studies support a general scheme for the behavior of the mammalian XY pair (Figure 1).

Each X and Y chromosome forms an axial filament (X or Y "axis") about 50 nm wide that has both ends attached to the nuclear envelope during leptotene or early zygotene. Both axes come close to each other at one end by late zygotene or early pachytene (the stage depending on the species) and from that end an SC begins to be assembled at early pachytene, reaching its greatest extent during the earlier pachytene substages. The SC then becomes progressively shorter, but the shortening is highly variable among species. Some species, like the Chinese hamster[68] and Armenian hamster,[71] may not show this desynapsis up to the end of pachytene. Despite this variability, in each species there is an average, characteristic extent of the SC. At the end of pachytene, minimal SC extents can be observed. During diplotene, the axes disappear. The speed and pattern of axial disassembly are also species-specific. An end-to-end attachment is the main association between sex chromosomes from diplotene to metaphase I.

In most species, the region that forms the SC piece behaves like the autosomes: the axes are regular, without differentiations, and the surrounding chromatin has a regular packing. The remaining parts of the X and Y axes undergo species-specific "differentiations" which may consist of thickenings, bulgings, branchings, and other bizarre deformations. Thus, in many species it is easy to distinguish "differential" and pairing pieces in the sex chromosome axes. This scheme has many variants, some of which are discussed in Sections V through VII.

B. Chronology

Several methods have been used to establish the sequence of events during the long meiotic prophase. Use of the stages of the cycle of the seminiferous epithelium,[72] coupled with serial sectioning and three-dimensional reconstructions, has been extensively described for the XY pair of the mouse[13,50] and rat.[69] The addition of other markers of cell activity (centriolar separation, nucleolar size, etc.) has been described in detail for human spermatocytes.[73] A simpler method is the sequential sampling of spermatocytes during the first (puberal) wave of spermatogenesis.[70] The use of hydroxyurea for selective killing of cells in the S-phase of the cycle, thus establishing a gap that can be used as a marker, has been applied to mice spermatocytes.[74] More commonly, stepwise, easily recognizable changes in cell structures (e.g., nucleolar growth, lengthening of the chromosomal axes, or changes in the axes of the sex chromosomes) have been used to construct a continuous sequence of changes, mainly in spread preparations.[68,75] In this way, five pachytene substages have been recognized in mouse spermatocytes[75] whose frequencies can be translated into time durations[75] on the assumption of proportionality between frequencies and durations. This method relies on the accurate determination of end-points. For pachytene, these end-points are the completion of synapsis (beginning of pachytene) and the initiation of autosomal desynapsis (diplotene). On the same basis, substages of pachytene have been described in spermatocytes of the deer mouse.[76]

These methods have shown some variations in the sequence of changes of the XY pair among mammalian species. Thus, in the human XY pair, an SC is not established until early pachytene,[19,73] and at the same time it reaches its maximum length,[19,77] However, in the chinchilla[78] and deer mice,[79] synapsis of the XY pair is delayed, and the maximum SC length is reached about mid-pachytene. The XY pair of the rat is probably intermediate in this respect.[70]

C. Differentiations in the Axes

The pieces of the X and Y axes outside the pairing region often show species-specific deformations, which may consist of thickenings,[13,70] splittings,[13,80] bulgings,[68,71] branchings,[19,81] net-like body formation,[19,81] diffuseness,[68] and differential staining.[68] The large bulgings found in the Armenian[71] and Chinese hamster[68] have been described in detail using three-dimensional reconstructions[71] and surface spreading.[68] These differentiations undergo elaborate changes during pachytene and these changes form a logical temporal sequence when grouped according to increasing complexity in the structures.[68] Furthermore, these changes can be independently put into a temporal sequence when observed in thin sections and related to the advance of spermatogenesis.[71] The evolving changes in differentiations are similar in both the Armenian and Chinese hamsters except at late pachytene and diplotene. The X and Y axes are delayed in pairing, compared to autosomes, but the regions that will form the end points of the SC are closely apposed at the end of zygotene. An SC about half the length of the Y axis is established at early pachytene and lasts up to late pachytene without desynapsing.[68,71] The unpaired regions first develop thickenings, which evolve to large bulgings — two in the X axis and one in the Y axis. A cloud-like staining of the distal regions precedes a redistribution of the axial material as the bulgings are transformed into an elaborate branching of thin and convoluted axes.[68] Although, in the Chinese hamster, desynapsis could not be followed in spreads,[68] desynapsis in the Armenian hamster is clearly shown at early diplotene since the SC opens up at the telomeric attachment and the SC remains are found only in the proximal region.[71] In the Armenian hamster, the bulgings have a hollow appearance in their centers and chromatin is definitely excluded at these centers.[71] The formation of bulgings, and the more common thickening, in the unpaired axes indicates an accumulation of axial materials which is not seen in SCs. Since the axes presumably are located between sister chromatids,[7] the distance between these chromatids seems to be enlarged in the unpaired regions. Furthermore, the regular accumulation of these materials, possibly proteins,[82] suggests that a particular molecular environment exists in these regions despite the lack of transcriptional activity in the XY body. Furthermore, in some cells the differential regions of the X axis are able to form an SC-like structure at late pachytene.[13] The multistranded organization of the X and Y axes, compared to that of autosomal SCs, has been shown in the mouse and rat.[80]

D. Synapsis and Desynapsis

The extent of synapsis in the XY pair is species-specific.[7,22] Several factors are involved in the limitation of synapsis. Since axes are attached to the nuclear envelope, the unequal lengths of the X and Y axis result in a steric hindrance for full pairing. In species such as the Armenian-[71] and Chinese hamsters[68] and the deer mouse,[79] synapsis stretches between fixed points and there is no desynapsis until diplotene. In some other mammalian species having small Y chromosomes, such as the South American primate *Cebus apella*,[38] the SC may be very short, even at early pachytene, and desynapsis during pachytene may give the appearance of nonpairing X and Y axes, as in cattle.[83]

The maximum extent of synapsis (SC) has been measured in a few species.[70,84] In the human XY pair, the average extent of the SC is 1.33 μm,[77] corresponding to 25.5% of the Y axis, that is, covering the whole short arm and the pericentromeric region.[77] On the other hand, the *maximum* SC length was reported to reach 81.8% of the Y axis[77] in one case and up to 72% of the Y axis in another study.[84] Thus, maximum lengths of the SC pass well beyond the kinetochore of the Y chromosome and thus include a region of nonhomologous synapsis. Nonhomologous synapsis can also occur between the nonpairing ends of the X and Y axes, as shown in the XY pair of the chinchilla[78] and, less frequently, in the rat.[70]

Desynapsis of part of the maximum extent reached by the SC seems to be the rule in species having an X chromosome of the "original" size, as in the mouse,[13] rat,[70] and human[19] (In the human, however, there is an opposite view.[73]) The extent of desynapsis may be limited by the occurrence of recombination nodules (RNs), as is probably the case in the Armenian hamster.[71] Desynapsis in the XY pair typically occurs during mid- and late pachytene, when autosomes continue to be associated by SCs, and thus it is assumed to differ from the lapse of synapsis that occurs at diplotene.

V. CHIASMATA, RECOMBINATION NODULES (RNs), AND CROSSING OVER

While most mammals show an end-to-end association at late prophase and metaphase I, some eutherian mammals display a clear chiasma in the XY pair or in trivalents that include the sex chromosomes. A non exhaustive list of these species includes the European hamster (*Cricetus cricetus*),[85] the Chinese hamster (*C. griseus*),[85] the Armenian hamster (*C. migratorius*),[71] the Sitka deer mouse (*Peromyscus sitkensis*),[79] the field mouse *Akodon mollis*,[86] *Artibeus jamaicensis*, and other stenodermatine bats,[44] the gerbils *Gerbillus gerbillus* and *G. pyramidum*,[87] the mongoose *Herpestes sanguineus*,[88] the musk shrew *Suncus murinus*,[89] and the rodent *Deltamys kempi*.[90] These species can be grouped into two classes: (1) species having X and Y chromosomes with added, C$^+$ heterochromatic segments and (2) species having a chromosomal rearrangement involving the X (or Y) chromosome and autosomes.

In the first group, except for the Armenian hamster, the chiasma occurs in a heterochromatic segment,[79,85] whereas, in the second group, the chiasma seems to originate in the autosomal segments included in the rearrangement, thus eliminating the necessity of the sex chromosomes to have their own mechanism of disjunction. Chiasmata seen in the first group are intriguing since heterochromatic regions seldom show recombination.[91] However, the specific regions involved in chiasma formation in the Chinese hamster seem to have special properties since they give little evidence of highly repetitive DNA sequences,[92] have intermediate resistance to alkali treatment, and are not stained by Giemsa-11.[93] There is little doubt that these chiasmata represent true recombination. In the Armenian hamster, the recombinational nature of the chiasma has been shown by the BrdU-dye method.[94] However, due to the usual organization of these sex-chromosome sets, they are not very helpful for a general analysis of the XY pair.

Recombination nodules have been recorded in the XY pair of a few mammals.[22] In the human XY pair, RNs have been located over the SC at early pachytene.[19,73] Most of the RNs are located near the telomeres, with 90% in the distal half of the short arm of the Y.[77] However, RNs are labile structures and difficult to preserve, and thus data on RNs are scarce.

Recently, some important advances have been made on the behavior of the human XY pair by the use of DNA probes, which had been foreseen in speculative papers. Polani[95] and Burgoyne[96] raised the assumption of regular crossing over in the pairing segment of the human XY pair. Polani[95] argued that the short arm of the Y chromosome may share homologous genes with the distal region of the short arm of the X chromosome. Furthermore, he assumed that, besides ''regular'' crossing over, recombination could occur (at lower frequencies) beyond the homologous region.[95] In such low frequency, non symmetrical recombination, the testis-determining factor (TDF gene) could be occasionally transferred to the X chromosome. Gametes carrying this TDF-bearing, X chromosome could give rise to XX males, as previously proposed by Ferguson-Smith.[97] Burgoyne[96] assumed that recombination in the human XY pair could be ''obligatory'' for a short, distal segment. Thus, genes situated beyond the location of the obligatory recombination would show no sex linkage, having a ''pseudoautosomal'' inheritance pattern. Most of these predictions have now been proved.

Homology between the distal regions of the short arms of the X and Y chromosomes in the human is supported by the regular formation of a SC.[19,37] This is further supported by their similar early-replication pattern.[60] The first gene to be mapped in both the X and Y human chromosomes has been the MIC2 gene, which codes for the cell-surface antigen 12E7.[98] This gene is distal to the TDF gene (located proximally in the Y chromosome), but is proximal with regard to the sequences showing moderate to complete recombination frequencies.[99] The cytological location of MIC2 is Xp22.3 and Yp11.2.[98] The recombination frequency of MIC2 is about 2.5%.[99] Besides gene (functional) homology, DNA - sequence homology in the X and Y chromosomes has been proved for a number of sequences, including an apparently telomeric one.[100] The DNA sequences shared by the tips of the short arms of the human X and Y chromosomes can be tentatively arranged in the following sequence, from the pairing telomeric region (Tel.) toward the centromeric region (Cen.) of the Y chromosome:[99] (Tel.) — DXYS14 —DXYS15 — DXYS17 — MIC2 — DXYZ2prox — TDF — (NTRS) — (Cen.)

The DNA probes used to recognize homologous sequences in the human X and Y chromosomes showed that some of these sequences were polymorphic with regard to their sensitivity to several restriction endonucleases; that is, they showed restriction-fragment-length polymorphism (RFLP). These differences in fragment length are inherited traits that have helped to analyze the behavior of the pairing region. Thus, the sequence DXYS14, which is located within 20 kb pairs from the telomeres, shows RFLP.[100] The tracing of

In 25% of the spermatocytes, the short arms of the autosomal axes were nonhomologously paired and no association with the XY was present. Spermatogenesis blockage, presumably at the spermatocyte- and spermatid stages, was attributed to an interference with the normal X-chromosome inactivation caused by the physical association of the telomeric heterochromatin of the autosomes.[153] Other cases of spermatogenic blockage in man associated with chromosomal variants have been interpreted in a similar way, e.g., the association of an additional 21 axis with the XY body[154] and the autosome-XY body association in autosomal translocations.[155] In other sterile patients, however, these associations with the XY body are not so frequent[156,157] and, when present, are mainly restricted to late pachytene substages.[155] Thus, it has been suggested that associations with the XY pair are instances of heterosynapsis which have no deleterious effects.[157] A low frequency (about 4%) of autosome-XY body associations has been reported in normal human males.[156] However, the high frequency of these associations in many infertile individuals has not been explained by alternative hypotheses.

E. Current Assumptions on the Mechanism of Spermatogenic Impairment

The XY body is transcriptionally inactive during pachytene (see Section III.D), while autosomes are very active in transcription.[55] This inactivity of the single X chromosome in male meiotic cells is enigmatic since it suppresses the production line for many "housekeeping" enzymes coded in the X chromosome. Furthermore, both X chromosomes are active in mammalian oocytes and, thus, X-chromosome inactivation seems to be peculiar to male meiosis in mammals. Thus, it has been postulated that this inactivation is needed for the normal development of spermatogenesis[21] and that any interference with this inactivation could lead to a blockage of spermatogenesis.[21] As applied to mammals,[112] this hypothesis assumes that the attachment of autosomal chromatin to the XY body interferes with the inactivation process, thereby allowing the expression of some gene activity in the sex chromosomes. The products derived from that activity could be detrimental to the pachytene spermatocyte, as are the products of nonpermissible mutations.[112] There is not yet an experimental basis for this assumption, except for the reported observation of increased levels of X-coded enzymes in cells from the testes of sterile mice.[158] Despite some contradictory observations (see Section VIII.D), this hypothesis remains attractive since it relates together the basic processes of sex chromatin- and XY body formation (both characteristic of mammals) with a high sensitivity of mammalian spermatogenesis with regard to the presence of X-autosome translocations and other rearrangements. This sensitivity is not present in either sex in bird gametogenesis (see Section IX).

It has recently been shown that some chromosomal abnormalities that induce male sterility in mammals also disturb germ-cell development in the ovary.[159,160] These observations support the assumption that a more general mechanism underlies most of the cases of gametogenic failure associated with chromosomal abnormalities.[114] According to this view, there is a functional relationship between the absence of chromosome pairing and gametogenic breakdown.[113,114] This hypothesis states that the normal association of *pairing sites* in the homologs at pachytene allows normal segregation and spermiogenic development. When these "sites" are not "saturated" by meiotic pairing, they are able to upset meiotic and postmeiotic development with a probability that is proportional to the number of "unsaturated" sites.[113] This assumption predicts that univalent sex chromosomes should contribute with a given probability to the start of gametogenic breakdown and that the recovery of gametic products carrying specific chromosomes will be inversely proportional to the number of "unsaturated sites" carried by the specific chromosomes. The hypothesis also suggests that cells carrying univalent chromosomes could escape gametogenic breakdown by self-pairing of the univalent.[113] In agreement with this hypothesis, self-pairing has been found in a significant proportion of XO oocytes in mice (see other chapters in this book).

This hypothesis is also related to the mechanism of meiotic breakdown in interspecific hybrids. As cited in Section VIII.C, the failure of synapsis seems to lead to meiocyte degeneration in hybrids. Thus, a mechanism might have been selected for the elimination of meiocytes in which synapsis does not reach some threshold level.

IX. THE AVIAN ZW PAIR

A. Morphology of the Z and W Chromosomes

Most avian species show a characteristic, heteromorphic sex pair (ZW) in the female sex.[24,161] In the usual gonosome set of birds, the W chromosome is medium to small in size, completely heterochromatic, and C^+ banding.[162,163] However, the phylogenetically primitive ratites do not have a specialized W chromosome.[164] The morphology of the Z chromosome is variable among avian families and it has been assumed that this variation arises from rearrangements of a conserved element since the size of the Z chromosome relative to the set, is apparently constant.[24] Some of the evolutionary changes of the Z chromosome, especially in *Galliformes,* have been tentatively traced.[161] These changes mainly involve a pericentric inversion that changes the ancestral, acrocentric Z element into a submetacentric chromosome, and the addition of C^+ heterochromatin to the telomeres of the short arm in the inverted Z chromosome.[161]

In the chicken, the Z chromosome ranks fifth in relative size[165] and is almost metacentric, while the W chromosome is metacentric and stands out clearly from the smaller macro-chromosomes. One of the Z arms ends in a typical C^+ band.[165] The identity of this marker C^+ band has been traced in Z-autosome translocations. Although the chicken is the best known avian species, the description of its meiotic process has been mostly restricted to the homogametic (male) sex.[166,167]

B. Partial Synapsis of the Z and W Chromosomes

Although the early stages of female meiosis in the chicken have been known since D'Hollander's classical description,[168] the synaptic association of these chromosomes was only recently described,[169] disproving a long-held assumption on the nonpairing of the Z and W chromosomes.

Oogenesis in the fowl proceeds in a definite, sequential way.[170] Oocytes at leptotene are found from day 17 of incubation, and about the time of hatching most of the oocytes are at the pachytene stage.[169,170] When the pachytene oocytes are observed with the micro-spreading techniques, the ZW pair is easily recognized as a medium-sized heteromorphic SC[169] (Figure 4). The Z axis represents 6.2% of the haploid SC set, and the average length of the W axis is 4.5% of that set.[169]

However, the length of the Z axis varies significantly during pachytene, as it becomes adjusted and almost equal in length to the W axis.[169] This shortening of the Z axis is accompanied by the thickening of its telomeric, nonpairing end.[169] The SC formed by the ZW pair is considerably longer than those of the XY pairs in mammals, having an average length of 10.4 µm.[171] This SC elongates significantly during synaptic adjustment of the Z and W axes.[171] This SC shows characteristic, asymmetric twists that are formed exclusively by the Z axis.[169] These observations have suggested that a major part of the SC in the ZW pair corresponds to nonhomologous synapsis, and that homologous pairing is restricted to the region close to the paired telomeres.[171]

Unlike the mammalian XY pair, no heteropycnosis is observed in the ZW pair at pachytene.[169] The state of the ZW pair at diplotene is mostly unknown, although it can be recognized at the beginning of diplotene[169] and, much later, at the well-developed lampbrush stage.[172]

C. Recombination Nodules in the ZW Pair

Recently, the invariable presence of a recombination nodule in the SC of the ZW pair of

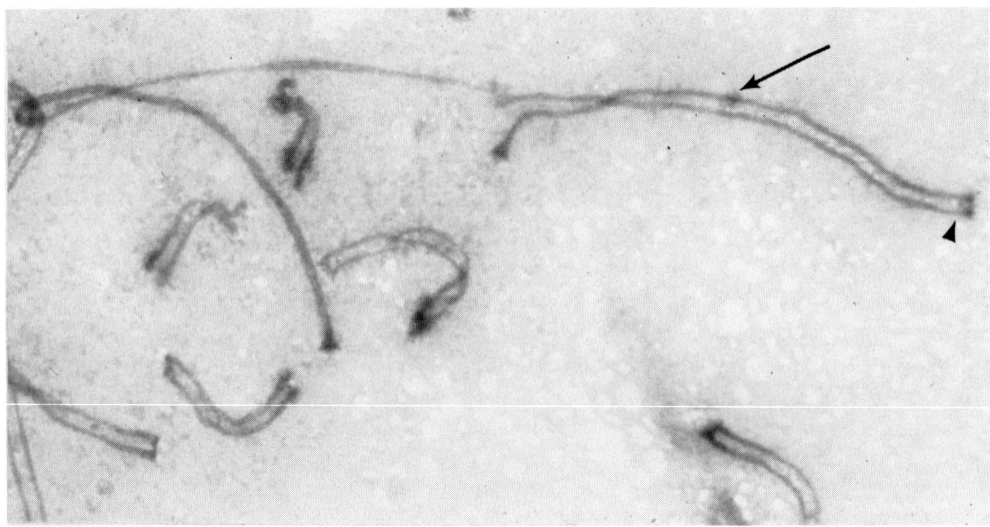

FIGURE 4.   ZW pair of chicken oocyte. The W axis is much shorter than the Z axis and both axes form a long SC, most of which is nonhomologous, as shown by the pairing of the W kinetochore (arrow) with the pairing arm of the Z axis. A dot-like recombination nodule (arrowhead) is found close to the paired telomeres of the Z and W axes. Microspread preparation. (Magnification × 10,300.)

the chicken has been described.[173] This nodule was located at an average distance of 0.31 μm from the paired telomeres and not more than 0.65 μm from them. Thus, this nodule shows a strict localization near the very tip of the paired region of the ZW pair.[173] This observation suggests that as in the human XY pair (see Section V), there is an obligatory recombination at the tip of the ZW pair.[173] In this case, the boundaries of the pseudoautosomal region should be more sharply defined since the variability of location of the nodules is much smaller in the chicken.[173] Furthermore, the strict localization of the nodule in the region close to the tip, supports the assumption that most of the length of the SC in the ZW pair represents nonhomologous pairing.

D. Abnormal Sex Complements

A wide variety of abnormal sex complements are viable and some of them are fully fertile in the chicken. In the chicken, triploid individuals are mainly ZWW (38%) and ZZZ (44%) at early embryogenesis,[174] but triploid adult birds are almost invariably ZZW in gonosomal constitution,[175,176] thus showing a sharp differential viability for the latter chromosomal constitution. ZZW triploid chickens are intersexes with ovotestis-like gonads that show only seminiferous histological structures, thus showing that the gonads are actually modified testes.[175] While the gonad of pure triploids is devoid of spermatozoa, diploid-triploid, mosaic birds can show 50% of the normal quantity of sperm in the gonad.[175] These observations correspond to the functional (left) gonad, the right gonad invariably having a testicular structure devoid of germ cells. The gametogenic blockage in triploid birds has not been thoroughly studied and in most cases the adult left gonad is reported to be lined with Sertoli cells only. However, some tubuli may have pachytene spermatocytes and even spermatids.[176] Meiotic studies in these triploids are limited to observations of the abnormal, triple association of chromosomal axes near the attachment to the nuclear envelope.[177] The joining of three lateral components of SCs has also been found in a few triploid or trisomic conditions (listed in von Wettstein et al.[22]). There is not yet sufficient information on differences in pairing patterns when more than two homologs are present, but observations on tetraploid chicken oocytes suggest that quadrivalent or bivalent formation is far more common than multiple pairing.[178]

Diploid-triploid mosaic chickens with ZZ/ZZZ, ZZ/ZZW, ZW/ZZZ, ZW/ZZW and ZW/ZWW gonosomal complements have been described at early embryogenesis.[179] Fertility of these mosaics depends on their sex-chromosome complements and the ratio of diploid/triploid cells,[180] while haploid-diploid mosaics having a 1AZ/2AZW constitution are normal, fertile females.[180]

Gonosome-autosome translocations in fowl have been extensively studied.[181-187] Unlike X-autosome translocations in mammals (see Section VIII), Z-autosome translocations do not quantitatively affect sperm production in cockerels.[186] Furthermore, the fertility of female carriers of a Z-microchromosome reciprocal translocation is not affected by the rearrangement.[183] Thus, the number of eggs laid shows no significant difference from that of normal hens, although there is a significant proportion of embryonic death before hatching due to genic imbalance.[183] Electron-microscopic studies both in males[187] and females heterozygous for Z-autosome translocations[188] have shown that pachytene meiocytes in these carriers show quadrivalents and other configurations with unpaired chromosomal axes, but no evidence of cell death or gametogenic impairment has been observed in these cases.

Thus, it may be concluded that meiosis in birds does not show the characteristic sensitivity toward sex-chromosome-autosome translocations, but meiosis can effectively be blocked by the presence of an additional chromosome set, as in triploid chickens.

X. CONCLUSIONS

The analysis of SC behavior by electron-microscopic techniques and, very recently, the application of DNA probes, have been leading the studies in this field until now. Some questions about the pairing of sex chromosomes have received fairly clear answers. Thus, partial synapsis of the heteromorphic chromosomes is an established rule for most eutherian mammals, as it probably is for birds. The chiasmatic nature of the end-to-end attachment of the human and mouse X and Y chromosomes at metaphase I is well-supported by the available evidence. The structural condensation of the XY body is expressed as a virtual abolition of transcription in most of this body. Spermatogenesis (and probably oogenesis as well) in mammals is especially sensitive to the presence of sex chromosome-autosome translocations, while this is not the case in bird gametogenesis.

These answers do not disguise the fact that molecular mechanisms operating in normal and abnormal meiosis are virtually unknown. The basic question — why meiotic cells of interspecific hybrids should die — points to our ignorance of the special, stringent metabolic conditions that are associated with the building and function of the protein structure proper of these cells and the SC, as well as of the special events of recombination and the preparation for disjunction that also occur at that stage. It is hoped that the next steps toward the understanding of normal and abnormal meiosis will come from the combination of the structural and biochemical approaches. The molecular basis of inactivation of the XY body will be one of these targets; the metabolic condition of unpaired chromosomes at pachytene will be another, perhaps related, target.

XI. ADDENDUM

A. TDF Locus and the Human XY Pairing Region

Y-chromosomal DNA has been demonstrated in a number of XX males.[189] By using *in situ* hybridization, this Y-DNA has been located on the distal end of the short arm of an X chromosome in three XX males.[189] An analysis of the "pseudoautosomal" region in the sex chromosomes of two XX males showed different results. In one patient, the entire pseudoautosomal region of the paternal Y chromosome had been transferred to an X chromosome, suggesting a reciprocal crossover between the X and Y of the patient's father; in the second

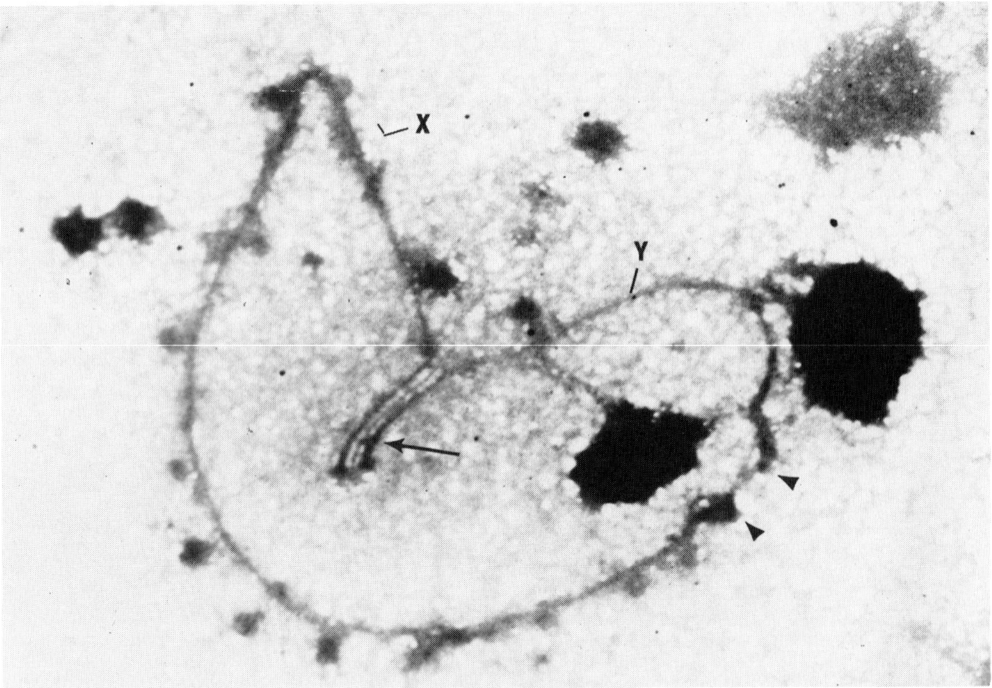

FIGURE 5.   Human XY pair with a recombination bar (arrow) on the near-terminal region of the SC. The distal telomeres of the X and Y axes are marked by arrowheads. (Magnification × 18,000.)

patient, there is evidence that an unequal crossover had occurred in paternal meiosis.[190] These data support the assumption that the relatively high incidence of human XX males (1 in 20,000 [189]) is related to an X-Y interchange involving the TDF locus in the Y chromosome. A 230-kb segment of the human Y chromosome containing at least part of the putative TDF locus has been cloned and its sequence analyzed.[191] The nucleotide sequences of this assumed gene are able to code for a zinc-complexing protein having 13 structural "fingers" and it has been assumed that this could be a regulatory, DNA-binding protein.[191] These sequences have been found in several mammals and in birds. In the Y chromosome of the mouse, they are located at the sex-determining region.[191] A similar sequence was detected in the X chromosome.[191] An upper estimate of the size of the pseudoautosomal region of the human XY pair is 5000 kb of DNA, according to the analysis of human-rodent somatic cell hybrids containing deleted and translocated human X chromosomes.[192] A theoretical model of XY recombination predicts that in the region undergoing recombination, allelic differentiation between the X and Y will develop only if the recombination rate is similar to or smaller than the mutation rate and, thus, little differentiation is expected for most of the recombining region.[193]

Immunocytochemical labeling of kinetochores in human SC spreads has confirmed that synapsis extends beyond the Y kinetochore.[194] As mentioned in Section V, recombination nodules are mostly distributed on the distal half of the short arm of the Y axis (Figure 5).

B. Sex Chromosome Pairing in Heterogametic Females

XY females have been extensively studied in the wood lemming *(Myopus schisticolor).*[195] A mutant X chromosome (X*) has been postulated to suppress the male-determining effect of the Y chromosome.[196] Data on sex chromosome pairing in these and other lemmings is scarce. X*Y females have oocytes with an orthodox X*X* constitution at diakinesis.[195] It

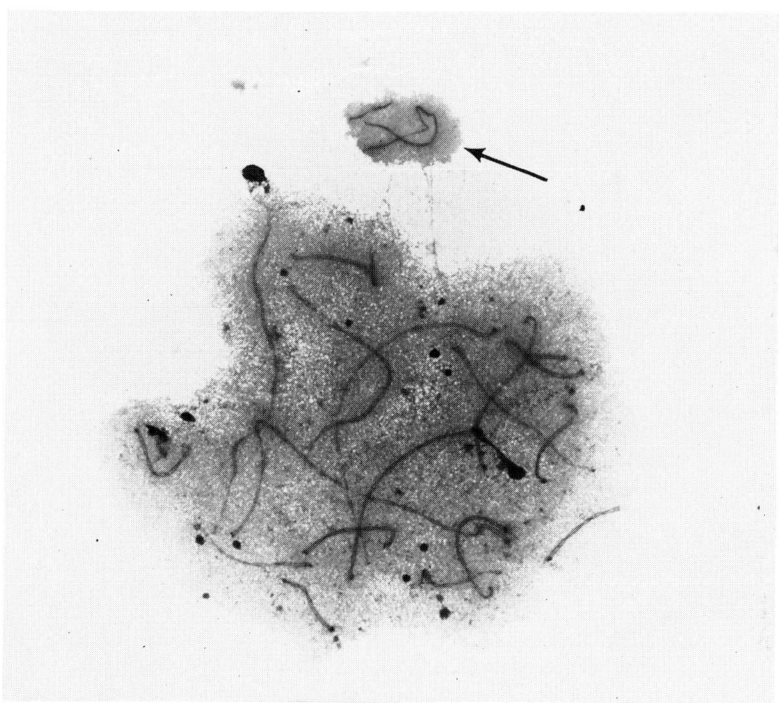

FIGURE 6.   Oocyte nucleus from an *Akodon azarae* female bearing a large deletion in one X chromosome. The heteromorphic sex chromosomes of these oocytes form a conspicuous "XY-like body" (arrow). (Magnification × 1500.)

has been assumed that nondisjunction in oogonia or oocytes results in the loss of the Y chromosome in germ cells.

The South American rodent *Akodon azarae* has a sex chromosome polymorphism that gives rise to different varieties of females.[197,198] Females carrying an extensive deletion of the long arm of the X chromosome have been bred in laboratory colonies and their offspring has been karyologically analyzed.[199] As mentioned in Section III.E, recent observations in oocytes from *Akodon* carrying the deleted X chromosome have shown a typical XY body ("sex vesicle") (Figure 6). The analysis of the axes involved in this body suggest that it is formed by the normal X and the deleted homolog, with a varied pattern of synapsis, mainly a nonhomologous one.[200]

C. Pairing in Mammalian XY Bodies

In the chiasmatic XY pair of the rodent *Micromys minutus,* synapsis is interstitially initiated at a proximal, euchromatic site: it then proceeds unidirectionally through the short arms, and finally forms a nonhomologous SC stretch along part of the long arm of the X axis.[201] Chiasmata occur only at the short arms. Strong evidence of the presence of regular SC formation in the XY pair of pure and hybrid cattle breeds has also shown that an artifactual, dissociated appearance of the X and Y axes may result from the breakage of the Y chromosome.[202] The possible link between XY autosome association and impaired spermatognesis is supported by the observation of XY quadrivalent contact in two subfertile human male carriers of different autosomal translocations.[203] The possible involvement of autosomal inactivation in such associations has been proposed.[204]

ACKNOWLEDGMENTS

I thank my collaborators and Dr. N. S. Fechheimer for their help while preparing this work. I am grateful for the hospitality of Ohio State University and, particularly, the Department of Dairy Sciences. This review was completed during my Visiting Professorship at that Department and with the support of a fellowship from CONICET. My personal thanks to Mrs. C. Daparci for her help in the preparation of most of the scientific materials.

REFERENCES

1. **Koller, P.C. and Darlington, C. D.,** The genetical and mechanical properties of the sex chromosomes. I. *Rattus norvegicus, J. Genet.,* 29, 159, 1934.
2. **Mittwoch, U.,** *Sex Chromosomes,* Academic Press, New York, 1967, chap. 9.
3. **Matthey, R.,** Les bases cytologiques de l'hérédité "relativement" liée au sexe chez les mammiferes, *Experientia,* 13, 341, 1957.
4. **Sachs, L.,** Sex-linkage and the sex chromosomes in man, *Ann. Eugen.,* 18, 255, 1954.
5. **Sachs, L.,** The possibilities of crossing-over between the sex chromosomes of the house mouse, *Genetica,* 27, 309, 1955.
6. **Ohno, S., Kaplan, W. D., and Kinosita, R.,** On the end-to-end association of the X and Y chromosomes of *Mus musculus, Exp. Cell Res.,* 18, 282, 1959.
7. **Solari, A. J.,** The behavior of the XY pair in mammals, *Int. Rev. Cytol.,* 38, 273, 1974.
8. **Moses, M. J.,** Chromosomal structures in crayfish spermatocytes, *J. Biophys. Biochem. Cytol.,* 2, 215, 1956.
9. **Solari, A. J.,** The morphology and the ultrastructure of the sex vesicle in the mouse, *Exp. Cell Res.,* 36, 160, 1964.
10. **Solari, A. J. and Tres, L.,** The ultrastructure of the human sex vesicle, *Chromosoma,* 22, 16, 1967.
11. **Solari, A. J.,** Changes in the sex chromosomes during meiotic prophase in mouse spermatocytes, *Genetics,* 61(Suppl.), 113, 1969.
12. **Moses, M. J.,** The synaptonemal complex and meiosis, in *Molecular Human Cytogenetics,* Sparkes, R. S., Comings, D. E., and Fox, C. F., Eds., Academic Press, New York, 1977, 101.
13. **Solari, A. J.,** The spatial relationship of the X and Y chromosomes during meiotic prophase in mouse spermatocytes, *Chromosoma,* 29, 217, 1970.
14. **Ford, E. H. R. and Woollam, D. H.,** The fine structure of the sex vesicle and sex chromosome association in spermatocytes of mouse, golden hamster and field vole, *J. Anat.,* 100, 787, 1966.
15. **Westergaard, M. and von Wettstein, D.,** The synaptonemal complex, *Ann. Rev. Genet.,* 6, 71, 1972.
16. **Solari, A. J.,** The behavior of chromosomal axes during diplotene in mouse spermatocytes, *Chromosoma,* 31, 217, 1970.
17. **Carpenter, A. T. C.,** Electron microscopy of meiosis in *Drosophila melanogaster* females. II. The recombination nodule — a recombination associated structure at pachytene?, *Proc. Natl. Acad. Sci. U.S.A.,* 72, 3186, 1975.
18. **Rasmussen, S. W. and Holm, P. B.,** Human meiosis. II. Chromosome pairing and recombination nodules in human spermatocytes, *Carlsberg Res. Commun.,* 43, 275, 1978.
19. **Solari, A. J.,** Synaptonemal complexes and associated structures in microspread human spermatocytes, *Chromosoma,* 81, 315, 1980.
20. **Ashley, T.,** Is crossover between the X and Y a regular feature of meiosis in mouse and man?, *Genetica,* 66, 161, 1985.
21. **Lifschytz, E. and Lindsley D. L.,** The role of X-chromosome inactivation during spermatogenesis, *Proc. Natl. Acad. Sci. U.S.A.,* 69, 182, 1972.
22. **Von Wettstein, D., Rasmussen, S. W., and Holm, P. B.,** The synaptonemal complex in genetic segregation, *Ann. Rev. Genet.,* 18, 331, 1984.
23. **White, M. J. D.,** *Animal Cytology and Evolution,* 3rd. ed., Cambridge University Press, Cambridge, 1973, chap. 16.
24. **Ohno, S.,** *Sex Chromosomes and Sex-linked Genes,* Springer-Verlag, Berlin, 1967, chaps. 1-4.
25. **Pathak, S. and Stock, A. D.,** The X chromosome of mammals: karyological homology as revealed by banding techniques, *Genetics,* 78, 703, 1974.

26. **Ohno, S., Beçak, W., and Beçak, M. L.,** X-autosome ratio and the behavior pattern of individual X-chromosomes in placental mammals, *Chromosoma,* 15, 14, 1964.
27. **Ohno, S., Stenius, C., Christian, L. C., Beçak, W., and Beçak, M. L.,** Chromosomal uniformity in the avian subclass Carinatae, *Chromosoma,* 15, 280, 1964.
28. **Ohno, S.,** Conservation *in toto* of the mammalian X-linkage group as a frozen accident, in *Chromosomes Today,* Vol. 9, Stahl, A., Luciani, J. M., and Vagner-Capodano, A. M., Eds., Allen & Unwin, Boston, 1987, 147.
29. **Baverstock, P. R., Adams, M., Polkinghorne, R. W., and Gelder, M.,** A sex-linked enzyme in birds: Z-chromosome conservation and lack of dosage compensation, *Nature,* 296, 763, 1982.
30. **Morizot, D. C., Bednarz, J. C., and Ferrell, R. E.,** Sex linkage of muscle creatine kinase in Harris' hawks, *Cytogenet. Cell Genet.,* 44, 89, 1987.
31. **Lyon, M. F.,** Mechanism and evolutionary origins of variable X-chromosome activity in mammals, *Proc. R. Soc. London Ser. B,* 187, 243, 1974.
32. **VandeBerg, J. L., Johnston, P. G., Cooper, D. W., and Robinson, E. S.,** X-chromosome inactivation and evolution in marsupials and other mammals, in *Isozymes: Current Topics in Biological and Medical Research,* Vol. 9, Rattazzi, M. C., Scandalios, J. G., and Whitt, G. S., Eds., Alan R. Liss, New York, 1983, 201.
33. **Murtagh, C. E.,** A unique cytogenetic system in monotremes, *Chromosoma,* 65, 37, 1977.
34. **Bull, J. J.,** Sex-determining mechanisms: an evolutionary perspective, *Experientia,* 41, 1285, 1985.
35. **Zenzes, M. T. and Wolff, V.,** Paarungsverhalten der Geschlechtschromosomen in der männlichen Meiose von *Microtus agrestis, Chromosoma,* 33, 41, 1971.
36. **Sharma, T.,** Germ cell chromosomes and their behavior during meiosis in a male Indian muntjak, *Muntiacus muntjak, Cytogenetics,* 11, 1, 1972.
37. **Solari, A. J. and Tres, L. L.,** The three-dimensional reconstruction of the XY chromosomal pair in human spermatocytes, *J. Cell Biol.,* 45, 43, 1970.
38. **Seuanez, H. N., Armada, J. L., Barroso, C., Rezende, C., and da Silva, V. F.,** The meiotic chromosomes of *Cebus apella* (Cebidae, Platyrrhini), *Cytogenet. Cell Genet.,* 36, 517, 1983.
39. **Solari, A. J. and Tres, L.,** The localization of nucleic acids and the argentaffin substance in the sex vesicle of mouse spermatocytes, *Exp. Cell Res.,* 47, 86, 1966.
40. **Howell, W. M. and Black, D. A.,** Controlled silver-staining of nucleolus organizer regions with a protective colloidal developer: a 1-step method, *Experientia,* 36, 1014, 1980.
41. **Dresser, M. E. and Moses, M. J.,** Synaptonemal complex karyotyping in spermatocytes of Chinese hamster *(Cricetulus griseus).* IV. Light and electron microscopy of synapsis and nucleolar development by silver staining, *Chromosoma,* 76, 1, 1980.
42. **Hulten, M., Solari, A. J., and Skakkebaek, N. E.,** Abnormal synaptonemal complex in an oligochiasmatic man with spermatogenic arrest, *Hereditas,* 78, 105, 1974.
43. **Liming, S. and Pathak, S.,** Gametogenesis in a male Indian muntjac × Chinese muntjac hybrid, *Cytogenet. Cell Genet.,* 30, 152, 1981.
44. **Tucker, P. K. and Bickham, J. W.,** Sex chromosome-autosome translocations in the leaf-nosed bats, family Phyllostomidae. II. Meiotic analyses of the subfamilies Stenodermatinae and Phyllostominae, *Cytogenet. Cell Genet.,* 43, 28, 1986.
45. **Raman, R. and Nanda, I.,** Identification and patterns of synapsis of the autosomically translocated Y-chromosome of the Indian mongoose, *Herpestes auropunctatus* (Hodgson), *Chromosoma,* 87, 477, 1982.
46. **Hogg, H. and McLaren, A.,** Absence of a sex vesicle in meiotic foetal germ cells is consistent with an XY chromosome constitution, *J. Embryol. Exp. Morph.,* 88, 327, 1985.
47. **Hayman, D. and Sharp, P.,** Verification of the structure of the complex sex chromosome system in *Lagorchestes conspicillatus* Gould (Marsupialia: Mammalia), *Chromosoma,* 83, 263, 1981.
48. **Solari, A. J.,** The behavior of chromosomal axes in Searle's X-autosome translocation, *Chromosoma,* 34, 99, 1971.
49. **Merry, D. E., Pathak, S., and VandeBerg, J. L.,** Differential NOR activities in somatic and germ cells of *Monodelphis domestica* (Marsupialia, Mammalia), *Cytogenet. Cell Genet.,* 35, 244, 1983.
50. **Solari, A. J.,** The evolution of the ultrastructure of the sex chromosomes (sex vesicle) during meiotic prophase in mouse spermatocytes, *J. Ultrastr. Res.,* 27, 289, 1969.
51. **Knibiehler, B., Mirre, C., Hartung, M., Jean, P., and Stahl, A.,** Sex vesicle-associated nucleolar organizers in mouse spermatocytes: localization, structure, and function, *Cytogenet. Cell Genet.,* 31, 47, 1981.
52. **Henderson, A. S., Eicher, E. M., Yu, M. T., and Atwood, K. C.,** The chromosomal location of ribosomal DNA in the mouse, *Chromosoma,* 49, 155, 1974.
53. **Kierszembaum, A. L. and Tres, L. L.,** Transcription sites in spread meiotic prophase chromosomes from mouse spermatocytes, *J. Cell Biol.,* 63, 923, 1974.
54. **Monesi, V., Geremia, R., D'Agostino, A., and Boitani, C.,** Biochemistry of male germ cell differentiation in mammals: RNA synthesis in meiotic and postmeiotic cells, *Curr. Topics Dev. Biol.,* 12, 11, 1978.

55. **Monesi, V.,** Synthetic activities during spermatogenesis in the mouse, *Exp. Cell Res.,* 39, 197, 1965.

56. **Speed, R. M.,** Abnormal RNA synthesis in sex vesicles of tertiary trisomic male mice, *Chromosoma,* 93, 267, 1986.

57. **De Boer, P. and Branje, H. E. B.,** Association of the extra chromosome of the tertiary trisomic male mice with the sex chromosomes during first meiotic prophase, and its significance for impairment of spermatogenesis, *Chromosoma,* 73, 369, 1979.

58. **Weisbrod, S.,** Active chromatin, *Nature,* 297, 289, 1982.

59. **Chandley, A. C. and McBeath, S.,** DNase I hypersensitive sites along the XY bivalent at meiosis in man include the XpYp pairing region, *Cytogenet. Cell Genet.,* 44, 22, 1987.

60. **Muller, U. and Schempp, W.,** Homologous early replication patterns of the distal short arms of pro-metaphasic X and Y chromosomes, *Hum. Genet.,* 60, 274, 1982.

61. **Weber, B., Schempp, W., and Wiesner, H.,** An evolutionarily conserved early replicating segment on the sex chromosomes of man and the great apes, *Cytogenet. Cell Genet.,* 43, 72, 1986.

62. **Venolia, L., Cooper, D. W., O'Brien, D. A., Millette, C. F., and Gartler, S. M.,** Transformation of the HPRT gene with DNA from spermatogenic cells, *Chromosoma,* 90, 185, 1984.

63. **Koffman-Alfaro, S. and Chandley, A. C.,** Meiosis in the male mouse. an autoradiographic investigation, *Chromosoma,* 31, 404, 1970.

64. **Zamboni, L. and Upadhyay, S.,** Germ cell differentiation in mouse adrenal glands. *J. Exp. Zool.,* 228, 173, 1983.

65. **Burgoyne, P. S., Levy, E. R., and McLaren, A.,** Spermatogenic failure in male mice lacking H-Y antigen, *Nature,* 320, 170, 1986.

66. **Solari, A. J. and Merani, S.,** unpublished data.

67. **Glamann, J.,** Crossing over in the male mouse as analyzed by recombination nodules, *Carlsberg Res. Commun.,* 51, 143, 1986.

68. **Moses, M. J.,** Synaptonemal complex karyotyping in spermatocytes of the Chinese hamster *(Cricetulus griseus).* II. Morphology of the XY pair in spread preparations, *Chromosoma,* 60, 127, 1977.

69. **Urena, F. and Solari, A. J.,** Three-dimensional reconstruction of the XY pair during pachytene in the rat *(Rattus norvegicus), Chromosoma,* 30, 258, 1970.

70. **Joseph, A. M. and Chandley, A. M.,** The morphological sequence of XY pairing in the Norway rat *Rattus norvegicus, Chromosoma,* 89, 381, 1984.

71. **Solari, A. J.,** The relationship between chromosomes and axes in the chiasmatic XY pair of the Armenian hamster *(Cricetulus migratorius), Chromosoma,* 48, 89, 1974.

72. **Oakberg, E. F.,** A description of spermiogenesis in the mouse and its use in analysis of the cycle of the seminiferous epithelium and germ cell renewal, *Am. J. Anat.,* 99, 391, 1956.

73. **Holm, P. B. and Rasmussen, S. W.,** Human meiosis. V. Substages of pachytene in human spermatogenesis, *Carlsberg Res. Commun.,* 48, 351, 1983.

74. **Dietrich, A. J. J. and de Boer, P.,** A sequential analysis of the development of the synaptonemal complex in spermatocytes of the mouse by electron microscopy using hydroxyurea and agar filtration, *Genetica,* 61, 119, 1983.

75. **Moses, M. J., Dresser, M. E., and Poorman, P. A.,** Composition and role of the synaptonemal complex, in *38th Symp. Soc. Exp. Biol.,* Controlling Events in Meiosis, Evans, C. W. and Dickinson, H. G., Eds., Company of Biologists, Cambridge, 1984, 245.

76. **Greenbaum, I. F., Hale, D. W., and Fuxa, K. P.,** The mechanism of autosomal synapsis and the substaging of zygonema and pachynema from deer mouse spermatocytes, *Chromosoma,* 93, 203, 1986.

77. **Solari, A. J.,** unpublished data.

78. **Solari, A. J. and Rahn, M. I.,** Asymmetry and resolution of the synaptonemal complex in the XY pair of *Chinchilla laniger, Genetica,* 67, 63, 1985.

79. **Hale, D. W. and Greenbaum, I. F.,** The behavior and morphology of the X and Y chromosomes during prophase I in the Sitka deer mouse *(Peromyscus sitkensis), Chromosoma,* 94, 235, 1986.

80. **Del Mazo, J. and Gil-Alberdi, L.,** Multistranded organization of the lateral elements of the synaptonemal complex in the rat and mouse, *Cytogenet. Cell Genet.,* 41, 219, 1986.

81. **Rosenman, A., Wahrman, J., Richler, C., Madgar, I., Weissenberg, R., and Chaki, R.,** Under what circumstances is the human XY bivalent tangled? A note on chromosomally derived sterility. *Cytogenet. Cell Genet.,* 45, 58, 1987.

82. **Solari, A. J.,** Ultrastructure and composition of the synaptonemal complex in spread and negatively stained spermatocytes of the golden hamster and albino rat, *Chromosoma,* 39, 237, 1972.

83. **Switonski, M., Gustavsson, I., and Ploen, L.,** The nature of the 1;29 translocation in cattle as revealed by synaptonemal complex analysis using electron microscopy, *Cytogenet. Cell Genet.,* 44, 103, 1987.

84. **Chandley, A., C., Goetz, P., Hargreave, T. B., Joseph, A. M., and Speed, R. M.,** On the nature and extent of XY pairing at meiotic prophase in man, *Cytogenet. Cell Genet.,* 38, 241, 1984.

85. **Vistorin, G., Gamperl, R., and Rosenkranz, W.,** Studies on sex chromosomes of four hamster species: *Cricetus cricetus, Cricetulus griseus, Mesocricetus auratus,* and *Phodopus sungorus, Cytogenet. Cell Genet.,* 18, 24, 1977.

86. **Semino, C., Oliveira, D., Bianchi, N. O., and Lobato, L.,** Cytogenetics of the South American akodon rodents (Cricetidae). IX. Chiasmatic sex bivalent in male meiosis of *Akodon mollis, J. Hered.,* 73, 149, 1982.

87. **Wahrman, J., Richler, C., Neufeld, E., and Friedmann, A.,** The origin of multiple sex chromosomes in the gerbil *Gerbillus gerbillus* (Rodentia: Gerbillinae), *Cytogenet. Cell Genet.,* 35, 161, 1983.

88. **Fredga, K.,** Unusual sex chromosome inheritance in mammals, *Philos. Trans. R. Soc. London Ser. B.* 259, 15, 1970.

89. **Satya-Prakash, K. L. and Aswathanarayana, N. V.,** Behavior of sex chromosomes during meiosis in the house shrew, *J. Hered.,* 75, 149, 1984.

90. **Sbalqueiro I. J., Mattevi, M. S., and Oliveira, L. F.,** An X1X1X2X2/X1X2Y mechanism of sex determination in a South American rodent, *Deltamys kempi* (Rodentia, Cricetidae), *Cytogenet. Cell Genet.,* 38, 50, 1984.

91. **John, B.,** Myths and mechanisms of meiosis, *Chromosoma,* 54, 295, 1976.

92. **Arrighi, F. E., Hsu, T. C., Pathak, S., and Sawada, H.,** The sex chromosomes of the Chinese hamster: constitutive heterochromatin deficient in repetitive DNA sequences, *Cytogenet. Cell Genet.,* 13, 268, 1974.

93. **Murer-Orlando, M. and Richer, C. L.,** Heterochromatin heterogeneity in Chinese hamster sex bivalents, *Cytogenet. Cell Genet.,* 35, 195, 1983.

94. **Allen, J. W.,** BrdU-dye characterization of late replication and meiotic recombination in Armenian hamster germ cells, *Chromosoma,* 74, 189, 1979.

95. **Polani, P. E.,** Pairing of X and Y chromosomes, non-inactivation of X-linked genes, and the maleness factor, *Hum. Genet.,* 60, 207, 1982.

96. **Burgoyne, P. S.,** Genetic homology and crossing over in the X and Y chromosomes of mammals, *Hum. Genet.,* 61, 85, 1982.

97. **Ferguson-Smith, M. A.,** X-Y chromosomal interchange in the aetiology of true hermaphroditism and of XX Klinefelter's syndrome, *Lancet,* II, 475, 1966.

98. **Buckle, V., Mondello, C., Darling, S., Craig, I. W., and Goodfellow, P. N.,** Homologous expressed genes in the human sex chromosome pairing region, *Nature,* 317, 739, 1985.

99. **Weissenbach, J.,** A molecular analysis of the human Y chromosome, in *Chromosomes Today,* Vol. 9, Stahl, A., Luciani, J. M., and Vagner-Capodano, A. M., Eds., Allen & Unwin, Boston, 1987, 165.

100. **Cooke, H. J., Brown, W. R. A., and Rappold, G. A.,** Hypervariable telomeric sequences from the human sex chromosomes are pseudoautosomal, *Nature,* 317, 687, 1985.

101. **Rouyer, F., Simmler, M. C., Johnsson, C., Vergnaud, G., Cooke, H. J., and Weissenbach, J.,** A gradient of sex linkage in the pseudoautosomal region of the human sex chromosomes, *Nature,* 319, 291, 1986.

102. **Singh, L. and Jones, K. W.,** Sex reversal in the mouse *(Mus musculus)* is caused by a recurrent nonreciprocal crossover involving the X and an aberrant Y chromosome, *Cell,* 28, 205, 1982.

103. **Evans, E. P., Burtenshaw, M. D., and Cattanach, B. M.,** Meiotic crossing-over between the X and Y chromosomes of male mice carrying the sex-reversing (Sxr) factor, *Nature,* 300, 443, 1982.

104. **Keitges, E., Rivest, M., Siniscalco, M., and Gartler, S. M.,** X-linkage of steroid sulphatase in the mouse is evidence for a functional Y-linked allele, *Nature,* 315, 226, 1985.

105. **Solari, A. J. and Ashley, T.,** Ultrastructure and behavior of the achiasmatic, telosynaptic XY pair of the sand rat *(Psammomys obesus), Chromosoma,* 62, 319, 1977.

106. **Solari, A. J. and Bianchi, N. J.,** The synaptic behavior of the X and Y chromosomes in the marsupial *Monodelphis dimidiata, Chromosoma,* 52, 11, 1975.

107. **Sharp, P.,** Sex chromosome pairing during male meiosis in marsupials, *Chromosoma,* 86, 27, 1982.

108. **Ratomponirina, C., Viegas-Pequignot, E., Dutrillaux, B., Petter, F., and Rumpler, Y.,** Synaptonemal complexes in Gerbillidae: probable role of intercalated heterochromatin in gonosome-autosome translocations, *Cytogenet. Cell Genet.,* 43, 161, 1986.

109. **Chandley, A. C.,** The genetics of human reproduction, *Experientia,* 42, 1109, 1986.

110. **De Boer, P.,** Chromosomal causes for fertility reduction in mammals, in *Chemical Mutagens,* Vol. 10, de Serres, F. J., Ed., Plenum Press, New York, 1986, 427.

111. **Magnuson, T. and Epstein, C. J.,** Genetic control of very early mammalian development, *Biol. Rev.,* 56, 369, 1981.

112. **Forejt, J.,** X-inactivation and its role in male sterility, in *Chromosomes Today,* Vol. 8, Bennett, M. D., Gropp, A., and Wolf, U., Eds., Allen & Unwin, London, 1984, 117.

113. **Miklos, G. L. G.,** Sex-chromosome pairing and male fertility, *Cytogenet. Cell Genet.,* 13, 558, 1974.

114. **Burgoyne, P. S. and Baker, T. G.,** Meiotic pairing and gametogenic failure, in *38th Symp. Soc. Exp. Biol., Controlling Events in Meiosis,* Evans, C. W. and Dickinson, H. G., Eds., Company of Biologists, Cambridge, 1984, 349.

115. **Ratcliffe, S. G., Stewart, A. L., Melville, M. M., Jacobs, P. A., and Keay, A. J.,** Chromosome studies on 3500 newborn male infants, *Lancet*, I, 121, 1970.

116. **Cattanach, B. M. and Pollard, C. E.,** An XYY sex-chromosome constitution in the mouse, *Cytogenetics*, 8, 80, 1969.

117. **Dobryanov, D. and Konstantinov, G.,** A case of mosaicism of the 58AXY, 58AXYY type in a male calf of the Bulgarian brown cattle breed, *C. R. Akad. Sci. Agric. Bulgaria*, 3, 271, 1970.

118. **Searle, J. B. and Wilkinson, P. J.,** The XYY condition in a wild mammal: an XY/XYY mosaic common shrew *(Sorex araneus), Cytogenet. Cell Genet.*, 41, 225, 1986.

119. **Hale, D. W. and Greenbaum, I. F.,** Spontaneous occurrence of XYY primary spermatocytes in the Sitka deer mouse, *J. Hered.*, 77, 131, 1986.

120. **Skakkebaek, N. E., Hulten, M., Jacobsen, P., and Mikkelsen, M.,** Quantification of human seminiferous epithelium. II. Histological studies in eight 47, XYY men, *J. Reprod. Fertil.*, 32, 391, 1973.

121. **Melnyk, J., Thompson, H., Rucci, A. J., Vanasek, F., and Hayes, S.,** Failure of transmission of the extra chromosome in subjects with 47,XYY karyotype, *Lancet*, II, 797, 1969.

122. **Evans, E. P., Beechey, C. V., and Burtenshaw, M. D.,** Meiosis and fertility in XYY mice, *Cytogenet. Cell Genet.*, 20, 249, 1978.

123. **Burgoyne, P. S.,** Evidence for an association between univalent Y chromosomes and spermatocyte loss in XYY mice and men, *Cytogenet. Cell Genet.*, 23, 84, 1979.

124. **Levy, E. R. and Burgoyne, P. S.,** The fate of XO germ cells in the testes of XO/XY and XO/XY/XYY mouse mosaics: evidence for a spermatogenesis gene on the mouse Y chromosome, *Cytogenet. Cell Genet.*, 42, 208, 1986.

125. **Burgoyne, P. S.,** The role of the sex chromosomes in mammalian germ cell differentiation, *Ann. Biol. Anim. Biochim. Biophys.*, 18, 317, 1978.

126. **Cattanach, B. M.,** XXY mice, *Genet. Res.*, 2, 156, 1961.

127. **Cattanach, B. M., Pollard, C. E., and Hawkes, S. G.,** Sex reversed mice: XX and XO males, *Cytogenetics*, 10, 318, 1971.

128. **McLaren A. and Monk, M.,** Fertile females produced by inactivation of an X chromosome of ''sex reversed'' mice, *Nature*, 300, 446, 1982.

129. **Eicher, E. M., Beamer, W. G., Washburn, L. L., and Whitten, W. K.,** A cytogenetic investigation of inherited true hermaphroditism in BALB/cWt mice, *Cytogenet. Cell Genet.*, 28, 104, 1980.

130. **McLaren, A., Simpson, E., Tomonari, K., Chandler, P., and Hogg, H.,** Male sexual differentiation in mice lacking H-Y antigen, *Nature*, 312, 552, 1984.

131. **Evans, E. P., Ford, C. E., and Searle, A. G.,** A 39,X/41,XYY mosaic mouse, *Cytogenetics*, 8, 87, 1969.

132. **Cattanach, B. M.,** Information on the Y chromosome and sex determination from mutants and other variants of sexual development, *Arch. Anat. microsc. Morphol. Exp.*, 74, 25, 1985.

133. **Winsor, E. J. T., Ferguson-Smith, M. A., and Shire, J. G. M.,** Meiotic studies in mice carrying the sex reversal (Sxr) factor, *Cytogenet. Cell Genet.*, 21, 11, 1978.

134. **Chandley, A. C. and Speed, R. M.,** Cytological evidence that the Sxr fragment of XY,Sxr mice pairs homologously at meiotic prophase with the proximal testis-determining region, *Chromosoma*, 95, 345, 1987.

135. **Gollapudi, B. B., Kamra, O. P., and Blecher, S. R.,** A search for a genetic basis for gonosomal univalence in mice, *Cytogenet. Cell Genet.*, 29, 241, 1981.

136. **Matsuda, Y., Imai, H. T., Moriwaki, K., and Kondo, K.,** Modes of inheritance of X-Y dissociation in inter-subspecies hybrids between BALB/c mice and *Mus musculus molossinus, Cytogenet. Cell Genet.*, 35, 209, 1983.

137. **Beatty, R. A., Lim, M.-C., and Coulter, V. J.,** A quantitative study of the second meiotic metaphase in male mice *(Mus musculus), Cytogenet. Cell Genet.*, 15, 256, 1975.

138. **Beechey, C. V.,** X-Y chromosome dissociation and sterility in the mouse, *Cytogenet. Cell Genet.*, 12, 60, 1973.

139. **Sotomayor, R. E. and Cumming, R. B.,** XY dissociation in mice: a model that may account for sex aneuploidy in humans, *Genetics*, 86, 60, 1977.

140. **De Boer, P. and Nijhoff, J. H.,** Incomplete sex chromosome pairing in oligospermic male hybrids of *Mus musculus* and *M. musculus molossinus* in relation to the source of the Y chromosome and the presence or absence of a reciprocal translocation, *J. Reprod. Fertil.*, 62, 235, 1981.

141. **Imai, H. T., Matsuda, Y., Shiroishi, T., and Moriwaki, K.,** High frequency of X-Y chromosome dissociation in primary spermatocytes of F1 hybrids between Japanese wild mice *(Mus musculus molossinus)* and inbred laboratory mice, *Cytogenet. Cell Genet.*, 29, 166, 1981.

142. **Searle, A. G. and Beechey, C. V.,** Sperm-count, egg-fertilization and dominant lethality after X-irradiation of mice, *Mutat. Res.*, 22, 63, 1974.

143. **Chaganti, R. S. K., Jhanwar, S. C., Ehrenbard, L. T., Kourides, I. A., and Williams, J. J.,** Genetically determined asynapsis, spermatogenic degeneration, and infertility in men, *Am. J. Hum. Genet.,* 32, 833, 1980.

144. **Vidal, F., Templado, C., Navarro, J., Brusadin, S., Marina, S., and Egozcue, J.,** Meiotic and synaptonemal complex studies in 45 subfertile males, *Hum. Genet.,* 60, 301, 1982.

145. **Chandley, A. C., Jones, R. C., Dott, H. M., Allen, W. R., and Short, R. V.,** Meiosis in interspecific equine hybrids. I. The male mule *(Equus asinus × E. caballus)* and hinny *(E. caballus × E. asinus), Cytogenet. Cell Genet.,* 13, 330, 1974.

146. **Pathak, S. and Kieffer, N. M.,** Sterility in hybrid cattle. I. Distribution of constitutive heterochromatin and nucleolus organizer regions in somatic and meiotic chromosomes, *Cytogenet. Cell Genet.,* 24, 42, 1979.

147. **Haldane, J. B. S.,** Sex ratio and unisexual sterility in hybrid animals, *J. Genet.,* 12, 101, 1922.

148. **Searle, A. G., Beechey, C. V., Evans, E. P., and Kirk, M.,** Two new X-autosome translocations in the mouse, *Cytogenet. Cell Genet.,* 35, 279, 1983.

149. **Ashley, T.,** Synaptonemal complex analysis of X-7 translocations in male mice, *Chromosoma,* 87, 149, 1982.

150. **Ashley, T.,** Nonhomologous synapsis of the sex chromosomes in the heteromorphic bivalents of two X-7 translocations in male mice: R5 and R6, *Chromosoma,* 88, 178, 1983.

151. **Forejt, J. and Gregorova, S.,** Meiotic studies of translocations causing male sterility in the mouse. I. Autosomal reciprocal translocations, *Cytogenet. Cell Genet.,* 19, 159, 1977.

152. **Forejt, J., Gregorova, S., and Goetz, P.,** XY pair associates with the synaptonemal complex of autosomal male-sterile translocations in pachytene spermatocytes of the mouse *(Mus musculus), Chromosoma,* 82, 41, 1981.

153. **Rosenmann, A., Wahrman, J., Richler, C., Voss, R., Persitz, A., and Goldman, B.,** Meiotic association between the XY chromosomes and unpaired autosomal elements as a cause of human male sterility, *Cytogenet. Cell Genet.,* 39, 19, 1985.

154. **Johannisson, R., Gropp, A., Winking, H., Coerdt, W., Rehder, H., and Schwinger, E.,** Down's syndrome in the male. Reproductive pathology and meiotic studies, *Hum. Genet.,* 63, 132, 1983.

155. **Gabriel-Robez, O., Ratomponirina, C., Dutrillaux, B., Carre-Pigeon, F., and Rumpler, Y.,** Meiotic association between the XY chromosomes and the autosomal quadrivalent of a reciprocal translocation in two infertile men, 46,XY,t(19;22) and 46,XY,t(17,21), *Cytogenet. Cell Genet.,* 43, 154, 1986.

156. **Chandley, A. C., Speed, R. M., McBeath, S., and Hargreave, T. B.,** A human 9;20 reciprocal translocation associated with male infertility analyzed at prophase and metaphase I of meiosis, *Cytogenet. Cell Genet.,* 41, 145, 1986.

157. **Batanian, J. and Hulten, M. A.,** Electron microscopic investigations of synaptonemal complexes in an infertile human male carrier of a pericentric inversion inv(1)(p32q42), *Hum. Genet.,* 76, 81, 1987.

158. **Hotta, Y. and Chandley, A. C.,** Activities of X-linked enzymes in spermatocytes of mice rendered sterile by chromosomal alterations, *Gamete Res.,* 6, 65, 1982.

159. **Burgoyne, P. S., Mahadevaiah, S., and Mittwoch, U.,** A reciprocal autosomal translocation which causes male sterility in the mouse also impairs oogenesis, *J. Reprod. Fertil.,* 75, 647, 1985.

160. **Mahadevaiah, S. and Mittwoch, U.,** Synaptonemal complex analysis in spermatocytes and oocytes of tertiary trisomic Ts(5^{12})31H mice with male sterility, *Cytogenet. Cell Genet.,* 41, 169, 1986.

161. **Belterman, R. H. R. and de Boer, L. E. M.,** A karyological study of 55 species of birds, including karyotypes of 39 species new to cytology, *Genetica,* 65, 39, 1984.

162. **Stefos, K. and Arrighi, F.,** Heterochromatic nature of W chromosome in birds, *Exp. Cell Res.,* 68, 228, 1971.

163. **Bloom, S. E.,** Current knowledge about the avian W chromosome, *Bioscience,* 24, 340, 1974.

164. **Takagi, N. and Sasaki, M.,** A phylogenetic study of bird karyotypes, *Chromosoma,* 46, 91, 1974.

165. **Pollock, D. L. and Fechheimer, N. S.,** Variable C-banding patterns and a proposed C-band karyotype in *Gallus domesticus, Genetica,* 54, 273, 1980.

166. **Pollock, D. L. and Fechheimer, N. S.,** The chromosomes of cockerels *(Gallus domesticus)* during meiosis, *Cytogenet. Cell Genet.,* 21, 267, 1978.

167. **Kaelbling, M. and Fechheimer, N. S.,** Synaptonemal complexes and the chromosome complement of domestic fowl, *Gallus domesticus, Cytogenet. Cell Genet.,* 35, 87, 1983.

168. **D'Hollander, F.,** Recherches sur l'oogenese et le noyau vitellin de Balbiani chez les oiseaux, *Arch. Anat. Microsc. Morphol. Exp.,* 7, 117, 1904.

169. **Solari, A. J.,** Ultrastructure of the synaptic autosomes and the ZW bivalent in chicken oocytes, *Chromosoma,* 64, 155, 1977.

170. **Hughes, G. C.,** The population of germ cells in the developing female chick, *J. Embryol. Exp. Morphol.,* 11, 513, 1963.

171. **Solari, A. J.,** unpublished data.

172. **Hutchison, N.,** Lampbrush chromosomes of the chicken, *Gallus domesticus, J. Cell Biol.,* 105, 1493, 1987.

173. **Rahn, M. I. and Solari, A. J.,** Recombination nodules in the oocytes of the chicken, *Gallus domesticus, Cytogenet. Cell Genet.,* 43, 187, 1986.

174. **Fechheimer, N. S.,** Origins of heteroploidy in chicken embryos, *Poultry Sci.,* 60, 1365, 1981.

175. **Abdel-Hameed, F. and Shoffner, R. N.,** Intersexes and sex determination in chickens, *Science,* 172, 962, 1971.

176. **De Boer, L. E. M., de Groen, T. A. G., Frankenhuis, M. T., Zonneveld, A. J., Sallevelt, J., and Belterman, R. H. R.,** Triploidy in *Gallus domesticus* embryos, hatchlings and adult intersex chickens, *Genetica,* 65, 83, 1984.

177. **Comings, D. E. and Okada, T. A.,** Triple chromosome pairing in triploid chickens, *Nature,* 231, 119, 1971.

178. **Solari, A. J. and Fechheimer, N. S.,** unpublished data.

179. **Fechheimer, N. S., Isakova, G. K., and Belyaev, D. K.,** Mechanisms involved in the spontaneous occurence of diploid-triploid chimerism in the mink *(Mustela vison)* and chicken *(Gallus domesticus), Cytogenet. Cell Genet.,* 35, 238, 1983.

180. **Thorne, M. H., Collins, R. K., and Sheldon, B. L.,** Live haploid-diploid and other unusual mosaic chickens, *Cytogenet. Cell Genet.,* 45, 21, 1987.

181. **Zartman, D. L.,** Location of the pea comb gene, *Poultry Sci.,* 52, 1455, 1973.

182. **Wooster, W. E., Fechheimer, N. S., and Jaap, R. G.,** Structural rearrangements of chromosomes in the domestic chicken: experimental production by X-irradiation of spermatozoa, *Can. J. Genet., Cytol.,* 19, 437, 1977.

183. **Telloni, R. V., Jaap, R. G., and Fechheimer, N. S.,** Fertility, embryo viability, and hatchability of chickens having 23% of the Z translocated to a micro-chromosome, *Poultry Sci.,* 56, 193, 1977.

184. **Dinkel, B. J., O'Laughlin-Phillips, E. A., Fechheimer, N. S., and Jaap, R. G.,** Gametic products transmitted by chickens heterozygous for chromosomal rearrangements, *Cytogenet. Cell Genet.,* 23, 124, 1979.

185. **Blazak, W. F. and Fechheimer, N. S.,** Gonosome-autosome translocations in the fowl: meiotic configurations and chiasma counts from singly and doubly heterozygous cockerels, *Can. J. Genet. Cytol.,* 22, 343, 1980.

186. **Blazak, W. F. and Fechheimer, N. S.,** Gonosome-autosome translocations in fowl: the development of chromosomally unbalanced embryos sired by singly and doubly heterozygous cockerels, *Genet. Res.,* 37, 161, 1981.

187. **Kaelbling, M. and Fechheimer, N. S.,** Synaptonemal complex analysis of chromosome rearrangements in the domestic fowl, *Gallus domesticus, Cytogenet. Cell Genet.,* 36, 567, 1983.

188. **Solari, A. J. and Fechheimer, N. S.,** unpublished data.

189. **Andersson, M., Page, D. C., and de la Chapelle, A.,** Chromosome Y-specific DNA is transferred to the short arm of X chromosome in human XX males, *Science,* 233, 786, 1986.

190. **Page, D. C., Brown, L. G., and de la Chapelle, A.,** Exchange of terminal portions of X- and Y-chromosomal short arms in human XX males, *Nature,* 328, 437, 1987.

191. **Page, D. C., Mosher, R., Simpson, E. M., Fisher, E. M. C., Mardon, G., Pollack, J., McGillivray, B., de la Chapelle, A., and Brown, L. G.,** The sex-determining region of the human Y chromosome encodes a finger protein, *Cell,* 51, 1091, 1987.

192. **Mondello, C., Ropers, H. H., Craig, I. W., Tolley, E., and Goodfellow, P. N.,** Physical mapping of genes and sequences at the end of the human X chromosome short arm, *Ann. Hum. Genet.,* 51, 137, 1987.

193. **Bengtsson, B. O. and Goodfellow, P. N.,** The effect of recombination between the X and Y chromosomes of mammals, *Ann. Hum. Genet.,* 51, 57, 1987.

194. **Sumner, A. T. and Speed, R. M.,** Immunocytochemical labelling of the kinetochore of human synaptonemal complexes and the extent of pairing of the X and Y chromosomes, *Chromosoma,* 95, 359, 1987.

195. **Fredga, K., Gropp, A., Winking, H., and Frank, F.,** Fertile XX- and XY-type females in the wood-lemming *Myopus schisticolor, Nature,* 261, 225, 1976.

196. **Bull, J. J. and Bulmer, M. G.,** The evolution of XY females in mammals, *Heredity,* 47, 347, 1981.

197. **Bianchi, N. O. and Contreras, J.,** The chromosomes of the field mouse *Akodon azarae* (Cricetidae, Rodentia) with special reference to sex chromosome anomalies, *Cytogenetics,* 6, 303, 1967.

198. **Bianchi, N. O., Dulout, F. N., and Contreras, J.,** Sex chromosome replication and sex chromatin in *Akodon azarae* (Rodentia Cricetidae), *Theor. Appl. Genet.,* 38, 343, 1968.

199. **Lizarralde, M. S., Bianchi, N. O. and Merani, M. S.,** Cytogenetics of South American Akodont rodents (Cricetidae). VII. Origin of sex chromosome polymorphism in *Akodon azarae, Cytologia,* 47, 183, 1982.

200. **Solari, A. J., Merani, M. S., Vitullo, A., and Agopian, S.,** Sex-chromosome behavior in oocytes of *Akodon azarae,* in preparation.

201. **Schmid, M., Johannisson, R., Haaf, T., and Neitzel, H.,** The chromosomes of *Micromys minutus* (Rodentia, Murinae). II. Pairing pattern of X and Y chromosomes in meiotic prophase. *Cytogenet. Cell Genet.,* 45, 121, 1987.

202. **Dollin, A. E. and Murray, J. D.,** personal communication.

203. **Johannisson, R., Lohrs, U., Wolff, H. H., and Schwinger, E.,** Two different XY-quadrivalent associations and impairment of fertility in men, *Cytogenet. Cell Genet.,* 45, 22, 1987.

204. **Ratomponirina, C., Couturier, J., Gabriel-Robez, O., Rumpler, Y., Dutrillaux, B., Croquette, M., Rabache, Q., and Leduc, M.,** Aberrations of the synaptonemal complexes in a male 46,XY, $-14, +$der(14)t(Y;14), *Ann. Genet.,* 28, 214, 1985.

Chapter 4

CHROMOSOME PAIRING AND FERTILITY IN PLANT HYBRIDS

Glyn Jenkins

TABLE OF CONTENTS

Table 1 (continued)
LIST OF HYBRIDS IN WHICH ZYGOTENE AND
PACHYTENE HAVE BEEN EXPLORED

Note: LM, conventional light microscopy of squash preparations; EM, electron microscopy of SCs of nonserial sections; 3D, three-dimensional reconstruction of SCs from serial electron micrographs; and SS, whole-mount surface spreading of SCs.

mosomes at pachytene in both sterile and fertile hybrids appears as intimate as that of the parental species. Similarly, in an F1 diploid hybrid between *Melilotus officinalis* and *M. alba*,[32] no structural differences between homoeologous chromosomes were evident at pachytene, but it suffered a 25% reduction in pollen fertility. In another hybrid from the same study, *M. messansensis* × *M. segetalis,* various structural differences at pachytene had no effect on bivalent formation at metaphase I, yet the hybrid is highly sterile. Clearly, in these cases, fertility is dependent not upon the intimacy or extent of chromosome synapsis, but rather upon the ability of structurally and genetically different homoeologs to form chiasmata and disjoin regularly at anaphase I.

It may be expected that very large differences in length between homoeologous chromosomes would disrupt pachytene pairing to such an extent that the fertility of the hybrid is reduced. However, two studies illustrate that this is not necessarily so. The two homoeologous chromosome complements of the F1 hybrids *Lycopersicon esculentum* × *Solanum lycopersicoides*[30] and *Crepis fuliginosa* × *C. neglecta*[8] differ in total length by 50 and 80%, respectively. At pachytene, however, these structural differences are virtually invisible. High pollen sterility in these two hybrids is the result of a reduction in chiasma frequency at metaphase I and not impairment of the synapsis of chromosomes.

Recent studies of chromosome pairing in three interspecific F1 hybrids, *Allium cepa* × *A. fistulosum, Festuca scariosa* × *F. drymeja,* and *Lolium temulentum* × *L. perenne,* warrant special attention, since they are particularly concerned with the inter- and intra-chromosomal structural differences and its effect upon chromosome pairing and fertility.

B. *Allium cepa* × *A. fistulosum*
1. Light Microscopy

Allium cepa and *A. fistulosum* are closely related, but differ by 27% in nuclear DNA amount, and by about 30% in total chromosome volume at metaphase of mitosis.[2,3] Despite this disparity, however, the two species hybridize readily. The F1 hybrid suffers reduced fertility, but by selecting appropriate parental genotypes, Maeda[40,41] has taken it to the F4 generation and has backcrossed it to the parents. At metaphase I in the hybrid, homoeologous chromosomes form chiasmata, albeit at a lower frequency than the parents, with univalents appearing in the majority of pollen mother cells (Figure 1). All the bivalents are heteromorphic and all the univalents of a pair are unequal in size, indicating that the DNA changes associated with the divergence of the two species, affected all the chromosomes of the complement. The degree of asymmetry is not the same in each bivalent, but varies in length from less than 20 to 60%. Clearly, the "extra" DNA belonging to *A. cepa* is not distributed in proportion to length. In fact, each chromosome of *A. cepa* has received approximately the same amount of extra DNA.[42]

The 27% difference in size between homoeologous chromosomes manifests itself at pachytene in the hybrid to a certain extent as unpaired loops and overlaps (Figure 2). These number one or two per configuration and suggest that the segmental duplications within the *A. cepa* chromosomes are highly localized.

2. Electron Microscopy
Recently, detailed observations have been made on pachytene pairing in this hybrid using

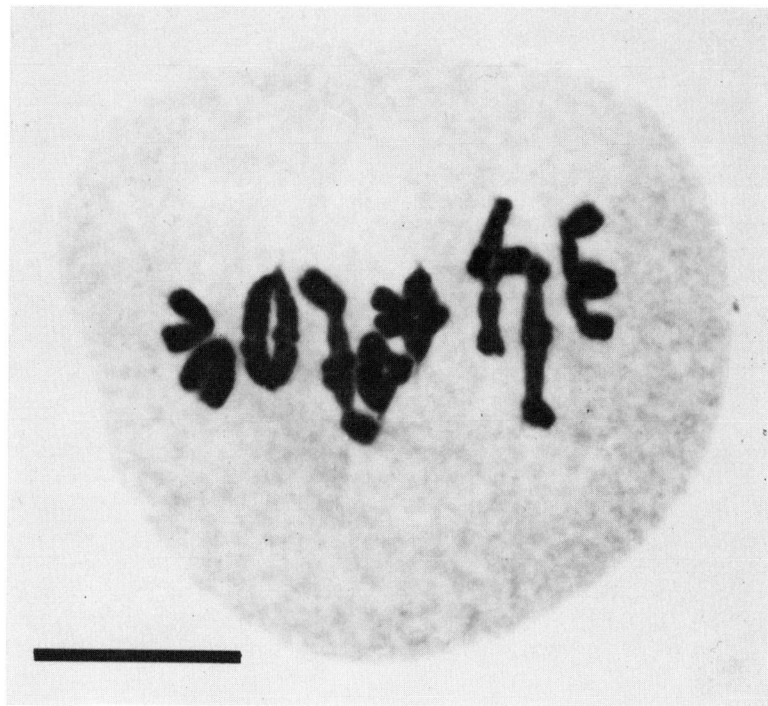

FIGURE 1.   A typical metaphase I plate in the hybrid *Allium cepa* × *A. fistulosum*. Note the pronounced heteromorphy in each bivalent and in the univalent pair. Bar represents 10 μm. (Courtesy of R. N. Jones.)

the relatively new technique of whole-mount surface spreading of synaptonemal complexes (SCs).[4] This method was originally developed by Counce and Meyer[43] in 1973 to study locust chromosomes and has since been adapted by several workers to reveal the SCs of plant cells. The technique employed in the study of the *Allium* hybrid essentially involves the enzymatic digestion of the callose wall surrounding the pollen mother cells, the bursting of the nuclei to release the chromosomes, the dispersion of the chromatin surrounding the SCs, and the drying down of the SCs onto plastic film which can be stained and transferred to the electron microscope.[44,45]

Albini[4] spread three nuclei that could be completely analyzed (Figure 3). Although the chromosomes of each were not fully paired, Albini considered the nuclei on morphological grounds to be at late pachytene. As expected, all the nuclei showed extensive two-by-two pairing between putative homoeologous chromosomes. In addition, however, nonhomologous SC formation was apparent, forming foldback loops within univalents and connecting heterologous chromosomes in multivalent configurations. Since associations of this nature do not appear at metaphase I, it was concluded that much of the SC formed during meiotic prophase is not effective in terms of supporting crossing over.

Homoeologous lateral components differed by between 5 and 20% in length. This difference is less than the 27% expected between parental DNA amounts, and considerably less than the 60% difference within some bivalents displaying large pachytene loops under the light microscope. Evidently, there is either some correction of the disparity in length at this stage or it may be that much of the "extra" DNA of *A. cepa* is not actually involved in SC formation. Although it was not possible to correlate the uneven distribution of length differences within the genome observed in surface spreads with those at metaphase I, it was speculated that the failure to pair or irregular SC formation may occur more frequently

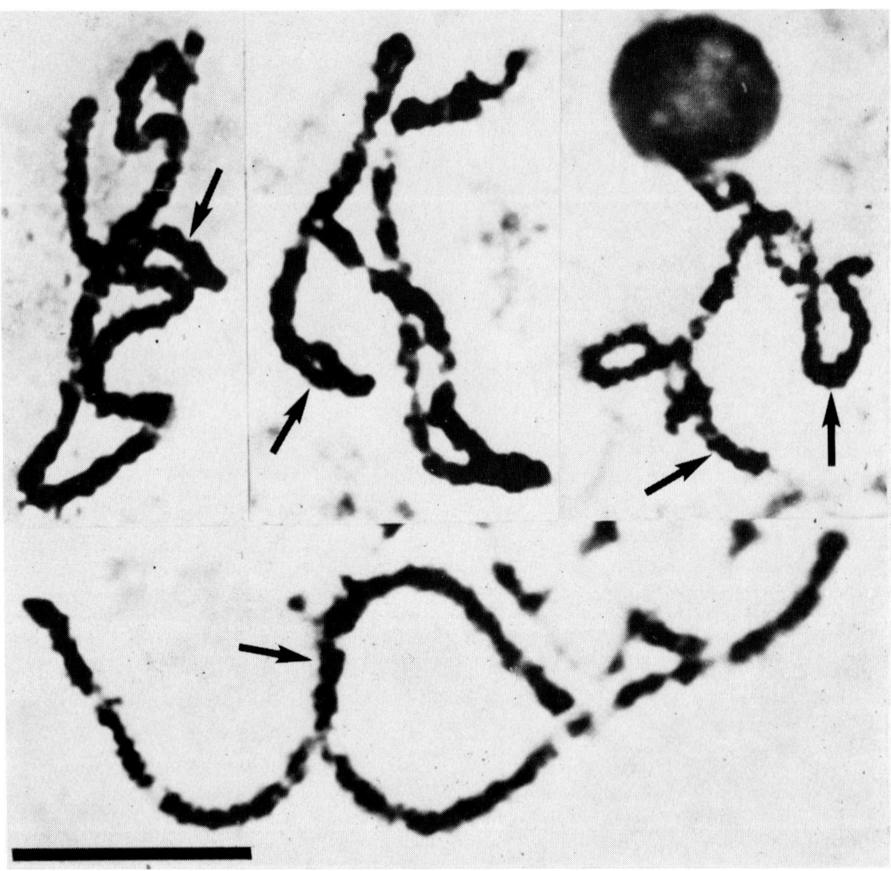

FIGURE 2.   Light micrographs of pachytene pairing configurations in the hybrid *Allium cepa* × *A. fistulosum*. Arrows indicate unpaired loops and overlaps. Bar represents 10 μm. (From Jones, R. N. and Rees, H., *Heredity*, 23, 591, 1968. With permission.)

between homoeologs which differ more markedly in size, or between those in which there is a pronounced localization of the "extra" DNA. Support for this notion, together with a detailed analysis of the accommodation of length discrepancies at the level of the SC, has come from a study of another hybrid, *Festuca scariosa* × *F. drymeja*.

C. *Festuca scariosa* × *F. drymeja*
1. Light Microscopy
The difference in nuclear DNA amount between *Festuca scariosa* and *F. drymeja* is 52%,[11,46] considerably greater than the 27% difference between the parental *Allium* species. In spite of this, the two fescues cross readily and produce viable hybrids. At metaphase I in the hybrid, however, there is a substantial reduction in chiasma frequency, with an average of only three bivalents and eight univalents per cell. Some cells were found, however, which had no more than two univalents and which could be used to observe the relative differences in size between chromosomes of homoeologous bivalents and univalent pairs. Estimation of relative volumes of homoeologs showed not only that all seven bivalents were asymmetrical, but also that the supplementary DNA of *F., drymeja* was distributed evenly between each member of the complement. In other words, each chromosome of *F. drymeja* carries approximately 0.2 pg more DNA than its *F. scariosa* homoeolog. A consequence of this distribution is that the smallest bivalents of the hybrid genome show the greatest degree of asymmetry.

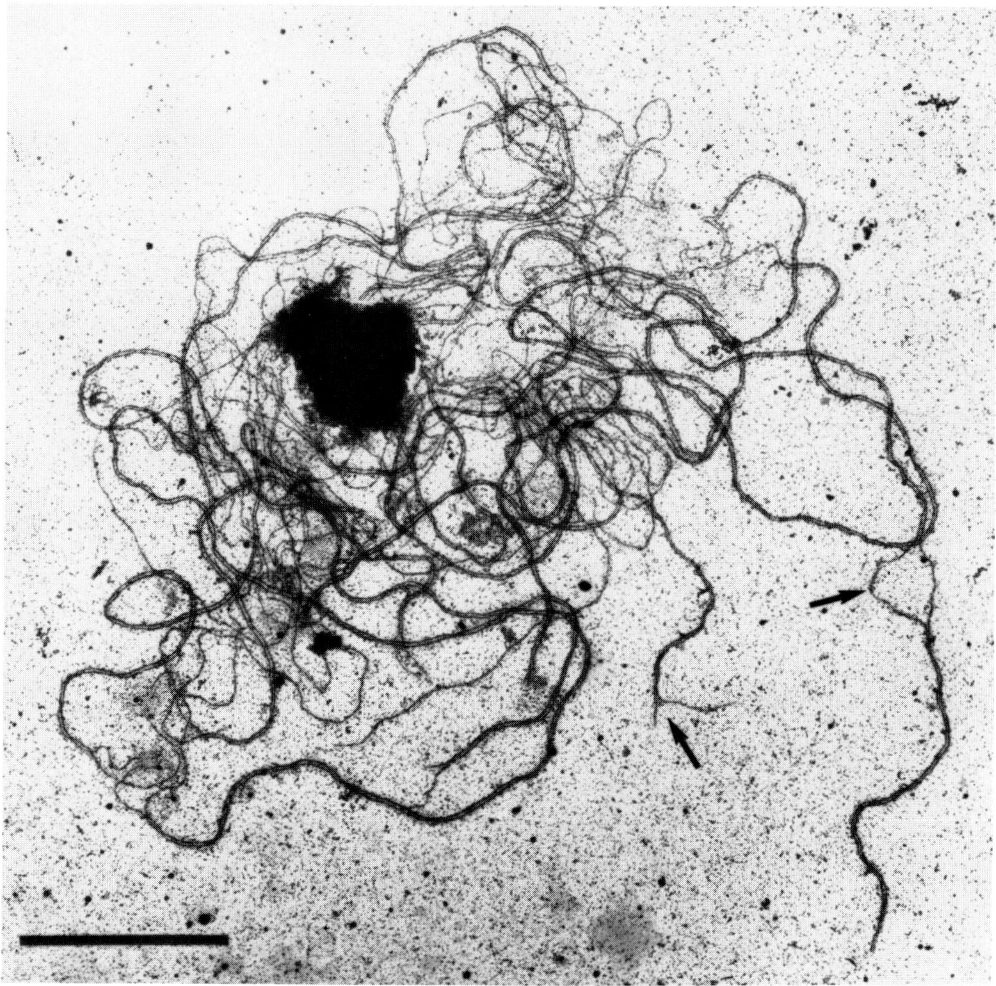

FIGURE 3. Electron micrograph of surface-spread pachytene SCs of a pollen mother cell nucleus of the hybrid *Allium fistulosum* × *A. cepa*. Note the disparity in lateral element lengths in an interstitial unpaired region and at a telomere (arrows). Bar represents 10 μm. (Courtesy of S. M. Albini.)

The failure of chiasmate association at metaphase I to some extent reflects failure of pairing during pachytene. Most cells at pachytene have three or four fully paired bivalents and a tangle of partially paired bivalents and univalents.[14] Clearly, in this case irregularity and failure of synapsis itself contributes significantly to the high pollen sterility of this hybrid. In spite of this hybrid being partially asynaptic, several cells were found which contained six or seven bivalents suitable for analysis (Figure 4). Since all the bivalents are heteromorphic at metaphase I, it was expected that discrepancies in length in the form of loops, buckles, and overlaps would be detected in all the bivalents of the complement at pachytene. Indeed, it was hoped that the positions of these loops would mark the sites of amplification of the duplicated segments belonging to *F. drymeja*. However, in the three largest bivalents of seven cells studied, there are no conspicuous pairing irregularities, despite differences of up to 40% in length. In contrast, the smaller bivalents fail to pair correctly from end to end. It may be that the greater relative differences in size between the chromosomes comprising the smaller bivalents is sufficient to disrupt the pairing process. It is not necessarily the only explanation, as noted in the following Section.

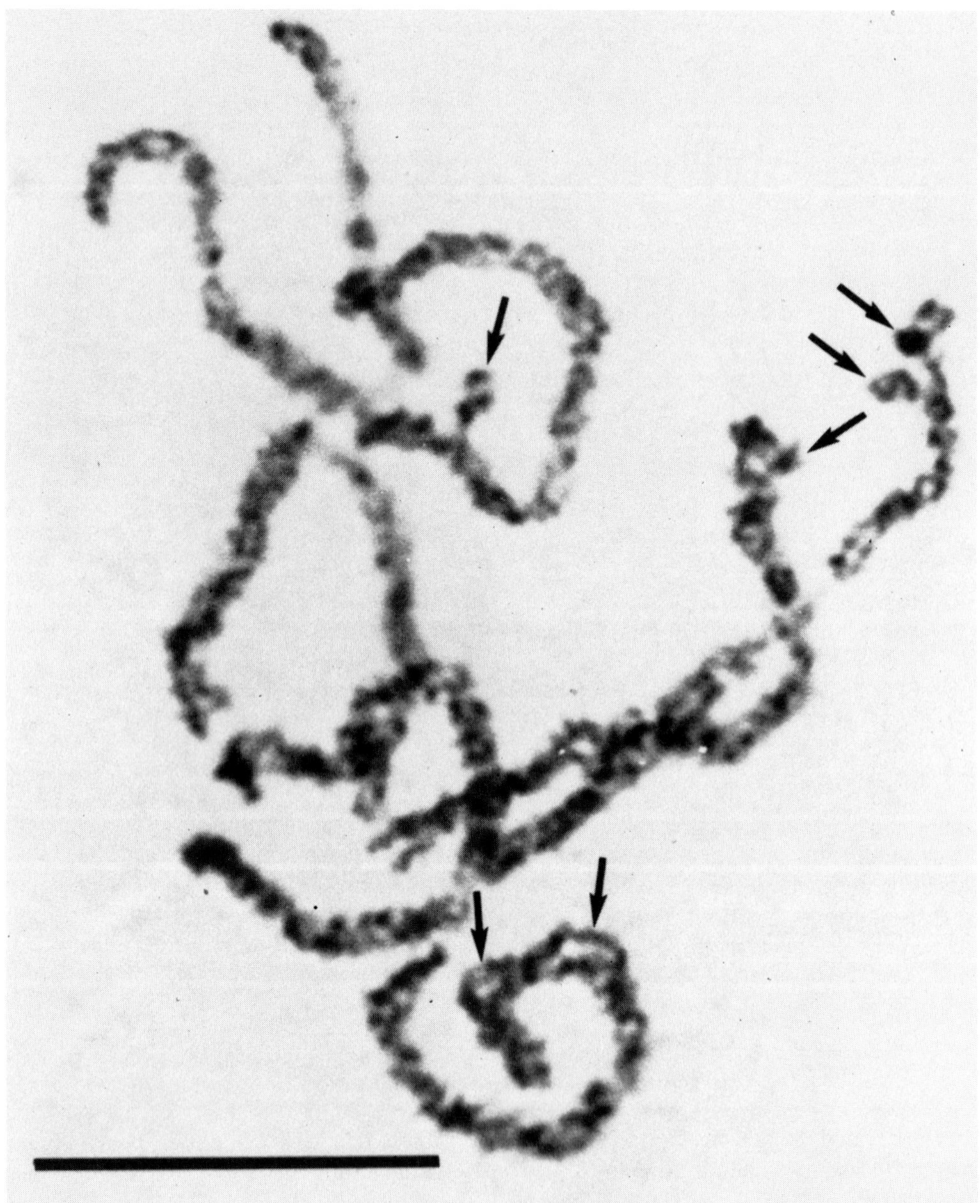

FIGURE 4.   Squash preparation of the seven homoeologous pachytene bivalents of the hybrid *Festuca scariosa* × *F. drymeja*. Arrows indicate unpaired loops. Note the remarkable completeness of pairing despite considerable length differences between homoeologs. Bar represents 10 μm. (From Seal, A. G., Ph.D. thesis, University College of Wales, Aberystwyth, 1981. With permission.)

2. Electron Microscopy

The inability to detect pachytene loops in the three largest bivalents prompted the use of a higher resolution technique, namely, three-dimensional reconstruction of SCs from serial electron micrographs.[25] In this method, consecutive ultra-thin serial sections are cut through pollen mother cell nuclei at zygotene or pachytene of meiosis. Each section is photographed in turn under the electron microscope and each electron micrograph assembled into a continuous series. SCs are identified and traced from one section to another until a complete three-dimensional picture of the chromosomes of each nucleus is built up. Computer transformation of the tracings to line diagrams enables identification of chromosomes both within and between nuclei. This technique preserves both the spatial distribution of the chromosomes within the nucleus and the integrity of the chromatin attached to the SCs.

In a study using this method,[15,16] 12 bivalents were fully reconstructed from six hybrid pachytene nuclei. By virtue of their large centromere volumes, they were identified as the larger bivalents of their respective nuclei.[47] This was particularly important since this set of bivalents constitutes those without loops when examined under the light microscope.[11] Careful inspection of these long bivalents showed that the SCs were apparently normal, complete, and uninterrupted from telomere to telomere. There were no unpaired lateral element loops to accommodate the expected length discrepancies. It could be that loops are formed initially, but are subsequently accommodated within the lateral elements of the SC by a process known as synaptic adjustment, as has been reported for the lateral elements of duplication heterozygotes in the mouse.[48] It is more likely, however, that the "extra" DNA does not contribute to the final length of the lateral elements of the larger homoeologs and does not participate in the formation of the SC. Since the supplementary DNA resides in areas outside the SC, and the SC effectively separates the chromatin of the two homoeologs of a bivalent (Figure 5), it was speculated that the sites of "extra" DNA could be determined simply by monitoring asymmetry of the chromatin about the SC. Within each of six reconstructed, fully paired bivalents it was possible to divide the SC into six regions of equal length and to estimate the relative volumes of chromatin of each homoeolog within each region (Figure 6). This facilitated a statistical analysis of the inter- and intrachromosomal distribution of the chromatin.[15] The test revealed that differences of up to 57% in total chromatin volume between homoeologs was indeed significant. Furthermore, the differences were not more pronounced in long or short arms, or in proximal, distal, or interstitial segments. In other words, the "extra" chromatin is spread randomly both within and between bivalents, with no similarities in distribution even between the same bivalents of different cells.

A similar analysis was performed on two medium-sized bivalents which were paired only in the long arms (Figure 6). In this case, the statistical test showed not only a significant difference in total chromatin volume between homoeologs, as in the fully paired bivalents, but also that there was a significant concentration of "extra" chromatin in the short arms. Since the total chromatin volume differences between the homoeologs of these two bivalents were no more than those of the fully paired bivalents, it was speculated that pairing failure in the short arms was due entirely to the uneven intrachromosomal distribution of the "extra" chromatin. Hence, the frequent complete failure of pairing of the smaller chromosomes might be the result not only of the relatively greater differences in size between homoeologs, but also of a highly localized distribution of "extra" material in *both* arms. In other words, the sterility of this hybrid is not the product of the magnitude of structural divergence of its homoeologs, but rather a result of the way in which the divergence has occurred.

Jenkins[15] and Jenkins and Rees[16] reconstructed all the chromosomes of one hybrid pachytene nucleus. In addition to some bivalent formation and widespread asynapsis, there was also extensive nonhomologous SC formation in the form of multivalents and foldback loops. To what extent these associations reflect homology between nonhomologous chromosomes is not known. However, the "extra" DNA of *F. drymeja* comprises, at least in part, repetitive sequences which have arisen or have been transposed throughout the entire genome. It is quite possible, therefore, that these sequences confer some degree of homology to chromosomes which are otherwise nonhomologous,[49] assuming, of course, that homology and the capacity to pair is determined simply on the grounds of structure. Rees et al.[12] speculated that in the parental species, which have no multivalents during meiotic prophase, the pairing between such sequences is under genetic control and is suppressed. In the hybrid, this control

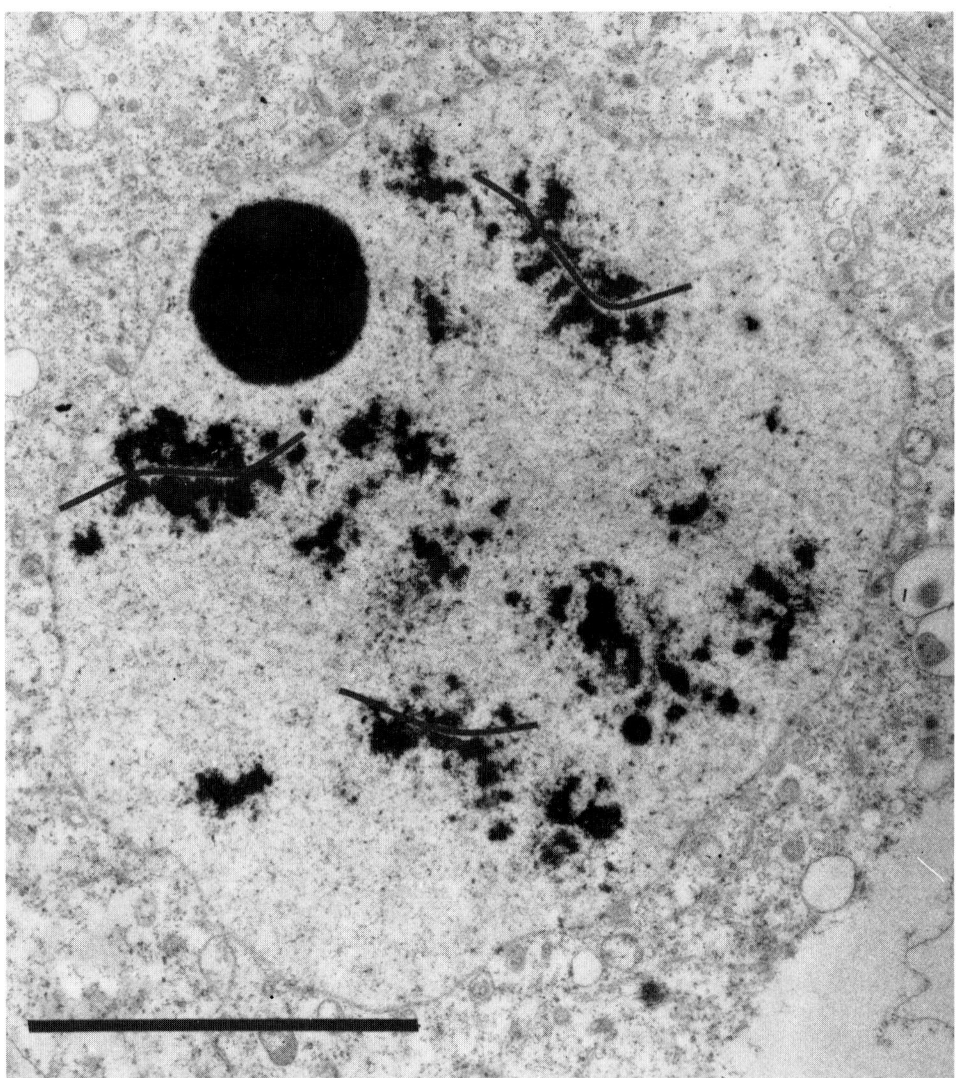

FIGURE 5.   Electron micrograph of a section through a pachytene nucleus of the hybrid *Festuca scariosa* × *F. drymeja* showing how the SC effectively separates the chromatin of each homoeolog within bivalents (lines). Bar represents 5 μm. (After Jenkins, G. and Rees, H., in *Kew Chromosome Conference II*, Brandham, P. E. and Bennett, M. D., Eds., Allen and Unwin, London, 1983, 233.)

breaks down, permitting pairing between nonhomologous chromosomes. It must be added, though, that despite structural similarity between these segments and a possible disruption of genotypic control in the hybrid, nonhomologous configurations do not support crossing over and do not persist to metaphase I.

D. *Lolium temulentum* × *L. perenne*
1. Light Microscopy

Pairing failure at pachytene in the *Festuca* hybrid was attributed both to the large disparity in size between the homoeologs of the smaller bivalents and to the intrachromosomal distribution of the supplementary DNA of *F. drymeja*. However, a study of chromosome pairing behavior in F1 hybrids between *Lolium temulentum* and *L. perenne* illustrates that these differences do not necessarily lead to widespread asynapsis. There is also an element of

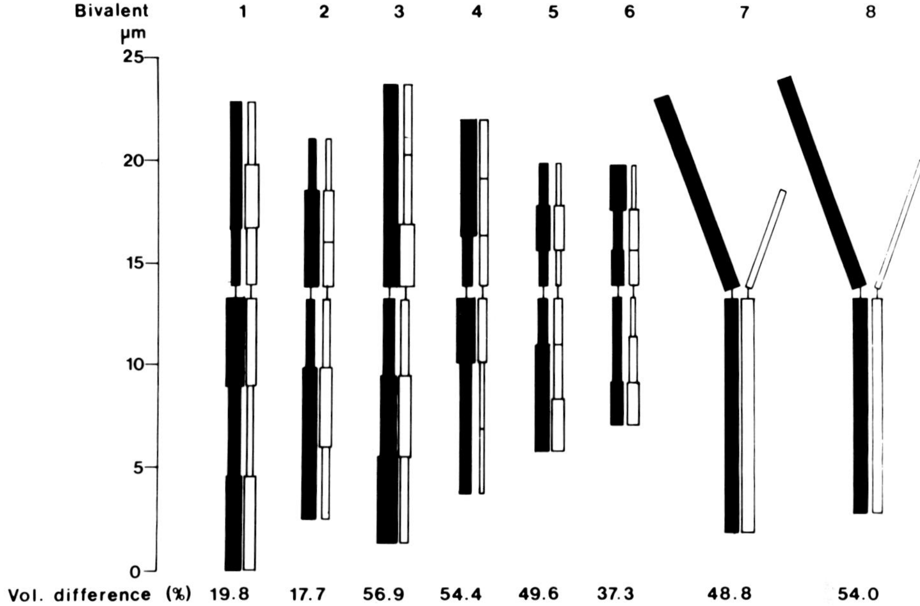

FIGURE 6.   Chromatin volume profiles of six reconstructed fully paired and two partially paired pachytene bivalents of the hybrid *Festuca scariosa* × *F. drymeja*. Because of similarities in morphology and total chromatin volume differences between homoeologs, bivalents 1 and 2, 3—5, and 7 and 8 were regarded as the same bivalents, respectively, of different nuclei. (After Jenkins, G. and Rees, H., in *Kew Chromosome Conference II*, Brandham, P. E. and Bennett, M. D., Eds., Allen and Unwin, London, 1983, 233.)

genotypic control of chromosome pairing and chiasma formation, which is hardly surprising considering its widespread occurrence within species of the Festuceae.[50]

L. temulentum and *L. perenne* differ in nuclear DNA amount by about the same degree as the parents of the fescue hybrid,[51] and the "extra" DNA of *L. temulentum* is apportioned roughly equally among the chromosomes of its complement,[11] again like the fescue hybrid. However, this hybrid has a chiasma frequency at metaphase I similar to one of its parents, with less than 20% of pollen mother cells containing univalents. Nevertheless, the F1 hybrid is male-sterile, but will produce progeny when backcrossed as female to *L. perenne*.[51] As expected, pairing at pachytene is regular and complete, with structural differences appearing as loops under the light microscope,[13,23,24] as in the hybrids of *Allium* and *Festuca*. Since all the chromosomes are paired, unlike in the *Festuca* hybrid, it was hoped to locate the sites of amplified DNA within the entire complement simply by mapping the intrachromosomal positions of loops. Unfortunately, this proved not to be possible under the light microscope because the chromosomes were too long and intertwined to allow a detailed comprehensive analysis.[11] It was necessary to employ a more sensitive technique, that of three-dimensional reconstruction of SCs, which would not only reveal the distribution of loops both within and between the chromosomes of the complement, but also the ultrastructure of the loops themselves.

2. Electron Microscopy

The SCs of five pachytene nuclei were reconstructed.[25] The mean karyotype is depicted in Figure 7. Despite an average DNA difference of 50% between homoeologs, 40% of the SCs run from telomere to telomere without loops, buckles, or overlaps. The remaining 60% of SCs form single lateral component loops which fold back upon themselves to form their own SC, which is indistinguishable from that of the main bivalent axis (Figure 8). The shorter bivalents have a significantly higher number of loops, as would be expected in view

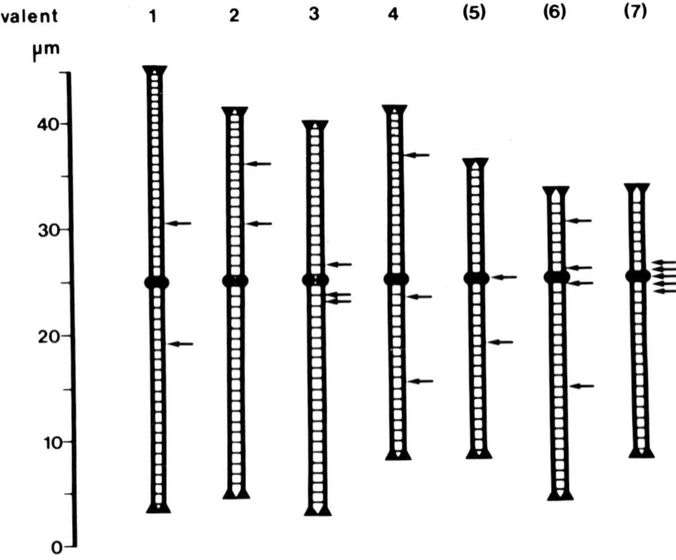

FIGURE 7.   The mean karyotype of the bivalents from five reconstructed pachytene nuclei of the diploid hybrid *Lolium temulentum* × *L. perenne*. SCs are represented by parallel vertical lines with cross-hatching, centromeres by filled circles, telomeres by filled triangles, and the sites of loops by arrows. On account of morphological similarity, bivalents 5—7 could not be positively identified. Note that the positions of the loops vary both within and between the bivalents of the complement. (After Jenkins, G., *Chromosoma*, 92, 81, 1985.)

of the greater asymmetry between homoeologs of the smaller bivalents. Even so, the loops account for only 14% of the expected difference in size between the two genomes of the hybrid. Coupled with the observation that many bivalents possess no loops at all, this indicates that the "extra" DNA is to a large extent not involved in SC assembly and is accommodated either prior to or during the assembly of the SC.

Figure 7 shows that loop position varies even within the same bivalent of different cells. This variation may be envisaged as a consequence of convergence of synapsis from opposite ends of the chromosomes, such that loops of unaccommodated lateral components arise at the point of convergence (Figure 9). Changes in the relative rates of synapsis from the telomeres would alter the point of convergence and the position of the loop and would inevitably incorporate nonhomologous segments of lateral components into the SC. The extent of heterosynapsis is difficult to ascertain since a nonhomologous SC is morphologically indistinguishable from its homologous counterpart. Nevertheless, the high incidence (67%) of mispaired centromeres indicates that nonhomologous SC formation occurs at least within the vicinity of centromeres and probably in other regions too. It should be emphasized, though, that both loop formation and accommodation appear to have little effect on the ability of the SC to support genetic exchange. Since chiasmata are distally located in this hybrid, it may be argued that homology is sufficiently well preserved in distal regions to permit crossing over.

Reconstruction of two zygotene nuclei revealed that chromosome pairing is by no means as regular as that seen at pachytene.[25] Pairing at this stage is permitted not only between homoeologous but also nonhomologous chromosome segments, resulting in the formation of bivalents, multivalents, and foldback loops. However, in view of the strict bivalent formation at pachytene, it must be concluded, as in the *Festuca* hybrid, that nonhomologous SC is illegitimate, ineffective in terms of chiasma formation, and will be eliminated in this

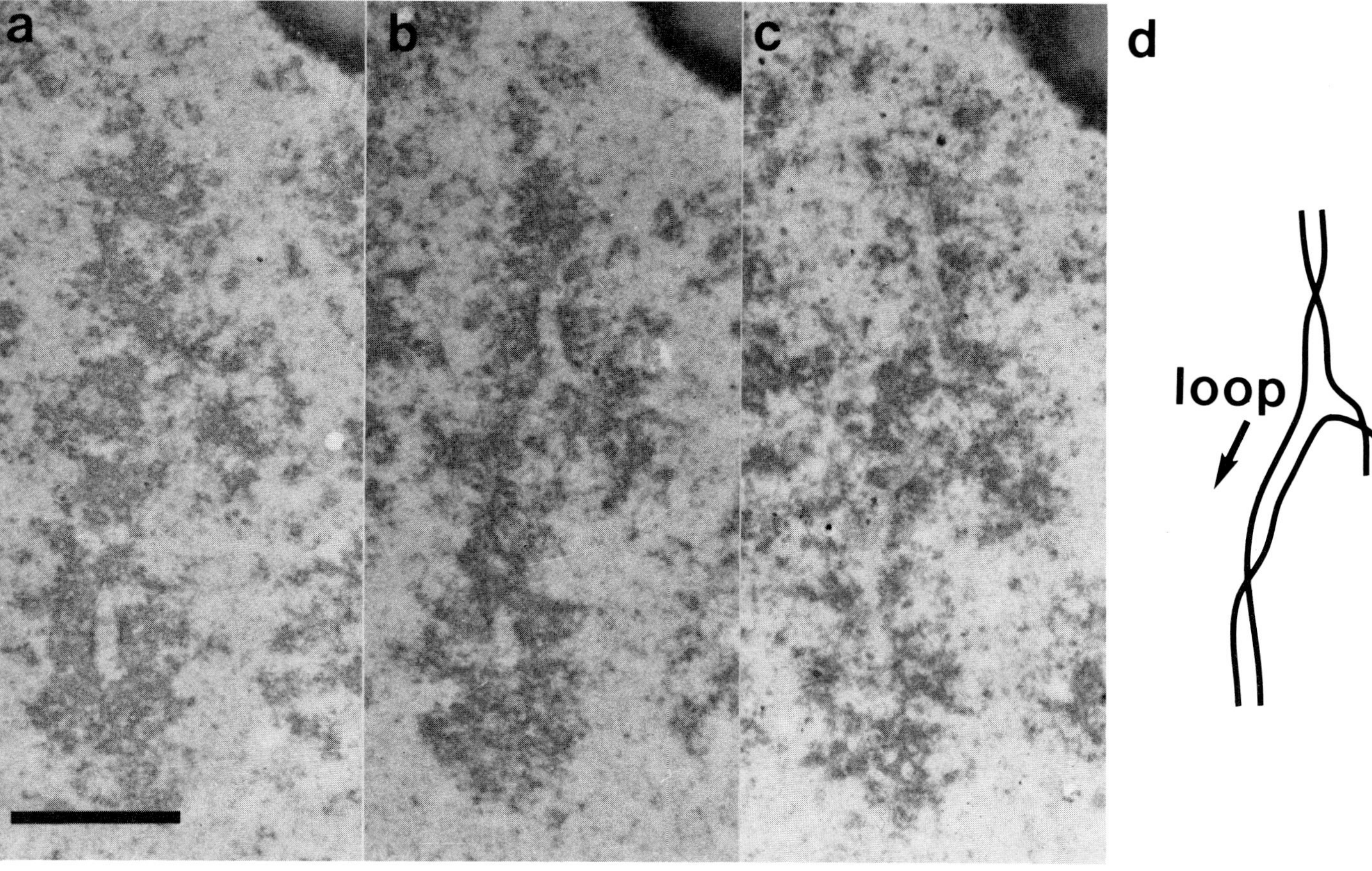

FIGURE 8. Consecutive electron micrographs (a—c) and interpretive drawing (d) through the point of emergence of a pachytene loop from the SC of a bivalent of the diploid hybrid *Lolium temulentum* × *L. perenne*. Note that the loop forms an SC identical with that of the main bivalent axis. Bar represents 1 μm. (After Jenkins, G., *Chromosoma* 92, 81, 1985.)

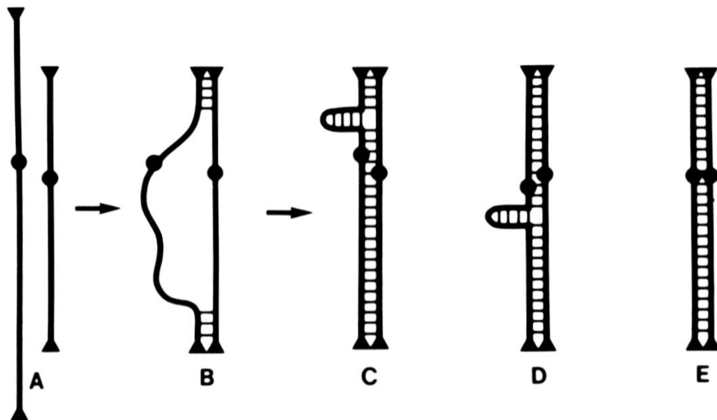

FIGURE 9.   Diagram to show how changes in the intrachromosomal positions of loops may arise. Symbols as in Figure 7. A pair of homoeologs of unequal length (A) begin pairing preferentially at the telomeres during zygotene (B). Depending on the relative rates of synapsis from the telomeres, loop positions will vary and centromeres will inevitably mispair (C and D). The disparity in length may be completely accommodated (E). (After Jenkins, G., in *Kew Chromosome Conference III*, in press.)

hybrid sometime before or during pachytene. The differences in length between homoeologous lateral components of bivalents at early zygotene are considerably less than the expected 50%, based on the difference in parental DNA amounts. This corroborates the observations at pachytene and shows that much of the ''extra'' DNA is never involved in lateral component assembly. Furthermore, the supplementary DNA that does not contribute, is subject to a process of accommodation throughout zygotene and perhaps pachytene also. For example, in one of the zygotene nuclei, bivalent 1 has lateral components which are only about 23% paired, yet show virtually no difference in length (Figure 10). Taken in isolation, it would be concluded that the ''extra'' DNA does not form a lateral component at all and the relative lengths are automatically equalized prior to synapsis. However, in an adjacent pollen mother cell, the same bivalent shows a 21% difference at approximately the same stage of pairing. Since bivalent 1 also forms loops at pachytene, it must be concluded that, to some extent, the ''extra'' DNA is involved in SC assembly, but that accommodation may start before and proceed throughout zygotene. It may even continue by synaptic adjustment thoughout pachytene[48] until all the loops have disappeared. Accommodation of length differences during zygotene and pachytene indicates that lateral components of the SC are quite flexible structures and can absorb large differences in length without adversely affecting the integrity (but not the homology) of the SC.

III. GENETIC CONTROL OF CHROMOSOME PAIRING

A. Genotypic Effect at Metaphase I

In the preceding section, it was shown that large discrepancies in size between homoeologs and the localization of amplified DNA within chromosomes can have a profound effect upon pachytene pairing, chiasma formation, and the fertility of hybrids. Structural difference is, however, only one of several factors that can markedly affect the chiasmate association of chromosomes. It has been known for some time, for example, that allohexaploid wheat contains genes that suppress the chiasmate associations of homoeologs at metaphase I.[52,53] The most important and potent of these reside at the Ph locus on the long arm of chromosome 5B, and are largely responsible for the cytologically diploid nature of this plant. Other classic allopolyploid species, such as *Festuca arundinacea* and *Avena sativa,* also form homologous

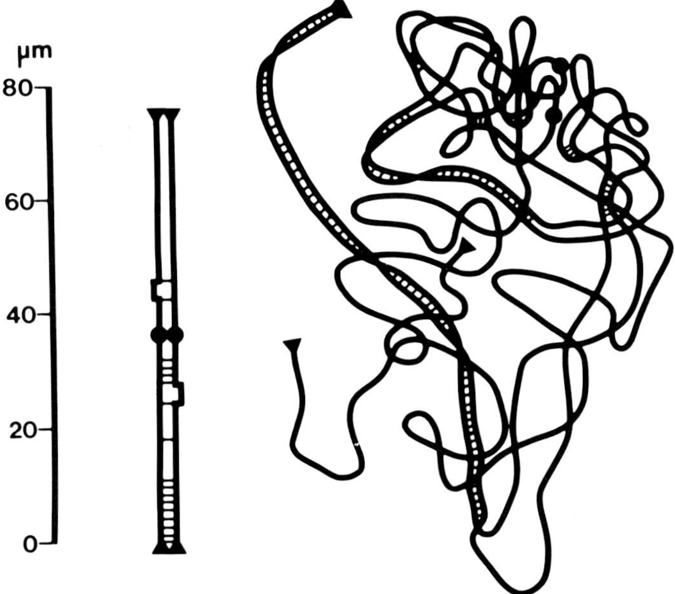

FIGURE 10. Original reconstruction and computer-derived diagram of the SC of a bivalent form a zygotene nucleus of the diploid hybrid *Lolium temulentum* × *L. perenne*. Symbols as in Figure 7. Note that even at only 23% pairing, homoeologous lateral components are virtually the same length.

bivalents only at first metaphase, although in these, the precise nature and location of the diploidizing genes are not known. In addition, accessory or supernumerary B chromosomes may influence chiasma frequencies and the association of homoeologs at metaphase I.[54] For example, in 1964, Mochizuki[55] showed that B chromosomes could compensate for the absence of the Ph locus by suppresssing chiasmate homoeologous association in a hybrid between nulli-5B *Triticum aestivum* × *Aegilops mutica*.

Table 2 lists a wide range of interspecific and intergeneric hybrids in which the effect of the genotype on chiasmate association at metaphase I has been studied. In some of the hybrids listed, the high degree of homoeologous association (+) is directly the result of the absence or inactivity of the Ph locus or its equivalent. In others, however, there are factors which actively inhibit the suppressors of homoeologous association and cause high levels of recombination between homoeologous chromosomes. The low degree of homoeologous association (−) of some hybrids in Table 2 may be due to the activity of Ph or Ph-like genes. With one exception, B chromosomes either have no detectable effect on chiasma frequencies or operate in a similar way to active Ph-like genes.

In some hybrids, particularly those of *Lolium,* a combined genic and B chromosome effect has been discovered. Chiasmate association between homoeologs of the hybrid *L. temulentum* × *L. perenne* is usually surprisingly high, considering the structural divergence of the two genomes. However, in the presence of B chromosomes, this association is very much reduced, resulting in a high number of univalents in a large proportion of the cells.[67,68,71] In the amphidiploid, B chromosomes have the same effect of suppressing chiasma formation between homoeologs, which results in a reduction in the number of multivalents and a promotion of homologous bivalent formation. A similar but less pronounced effect was described in a diploid hybrid between *L. multiflorum* and *L. perenne*.[63] At the tetraploid level, however, the B chromosomes appeared not to alter the pattern of association. It was suggested that this was because the two chromosome complements were structurally so similar[24] that the B chromosome mechanism could not discriminate between them.

Table 2

LIST OF HYBRIDS IN WHICH A PROMOTION (+), AN INHIBITION (−) OR NO EFFECT (0) ON HOMOEOLOGOUS CHROMOSOME ASSOCIATION AT METAPHASE I HAS BEEN OBSERVED UNDER THE INFLUENCE OF GENES AND B CHROMOSOMES

Hybrid	Gene	B	Ref.
Aegilops mutica × *Ae. caudata* (2x)		−	56
Ae. mutica × *Ae. comosa* (2x)		−	56
Ae. mutica × *Ae. longissima* (2x)		−	56
Ae. mutica × *Ae. speltoides* (2x)		−	56
Ae. speltoides × *Triticum araraticum* (3x)		−	57
Ae. speltoides × *T. dicoccoides* (3x)		−	57
Ae. speltoides × *T. dicoccum* (3x)		−	57
Ae. speltoides × *T. durum* (3x)		−	57
Avena longiglumis × *A. sativa* (4x)	+		58
Festuca arundinacea × *Lolium perenne* (4x)	−	−	59
F. pratensis × *L. multiflorum* (2x,3x)		0	60
F. pratensis × *L. multiflorum* (2x,3x)		0	61
L. multiflorum × *L. loliaceum* (2x)		0	62
L. multiflorum × *L. perenne* (2x,4x)		−	63
L. multiflorum × *L. perenne* (4x)	−		64
L. multiflorum × *L. persicum* (2x)		0	62
L. multiflorum × *L. remotum* (2x)		0	62
L. multiflorum × *L. strictum* (2x)		0	62
L. perenne × *F. arundinacea* (4x)		−	65
L. perenne × *F. pratensis* (2x,3x)		0	60
L. perenne × *F. pratensis* (2x,3x)		0	61
L. perenne × *L. multiflorum* (2x,3x)		0	60
L. perenne × *L. multiflorum* (2x,3x)		0	61
L. perenne × *L. multiflorum* (2x)		0	62
L. perenne × *L. rigidum* (2x)		0	62
L. rigidum × *L. loliaceum* (2x)		0	62
L. rigidum × *L. persicum* (2x)		0	62
L. rigidum × *L. remotum* (2x)		0	62
L. rigidum × *L. strictum* (2x)		0	62
L. rigidum × *L. temulentum* (2x)		0	62
L. temulentum × *L. multiflorum* (2x,3x,4x)	−		66
L. temulentum × *L. perenne* (2x,4x)		−	67
L. temulentum × *L. perenne* (2x,4x)		−	68
L. temulentum × *L. perenne* (2x)	−	−	69
L. temulentum × *L. perenne* (2x,4x)	−		70
L. temulentum × *l. perenne* (2x)		−	71
L. temulentum × *L. perenne* (4x)	−	−	72
L. temulentum × *L. perenne* (2x)	−		73
L. temulentum × *L. rigidum* (2x)	−	−	69
L. temulentum × *L. rigidum* (2x)	−	−	74
Triticale × *Secale cereale* (5x)	+		75
Triticale × *Triticum aestivum* (6x)	+		76
T. aestivum × *Ae. longissima* (4x)	+		77
T. aestivum × *Ae. longissima* (4x)	+		78
T. aestivum × *Ae. mutica* (4x)		−	55
T. aestivum × *Ae. mutica* (4x)		−	79
T. aestivum × *Ae. mutica* (4x)	+ / −[a]		80
T. aestivum × *Ae. mutica* (4x)		0	81
T. aestivum × *Ae. speltoides* (4x)	+		78
T. aestivum × *Ae. speltoides* (4x)	+		82
T. aestivum × *Ae. speltoides* (4x)		−	79
T. aestivum × *Ae. speltoides* (4x)		0	81

Table 2 (continued)
**LIST OF HYBRIDS IN WHICH A PROMOTION (+), AN
INHIBITION (−) OR NO EFFECT (0) ON
HOMOEOLOGOUS CHROMOSOME ASSOCIATION AT
METAPHASE I HAS BEEN OBSERVED UNDER THE
INFLUENCE OF GENES AND B CHROMOSOMES**

Hybrid	Gene	B	Ref.
T. aestivum × *S. cereale* (4x)		−	83
T. aestivum × *S. cereale* (4x)	+		84
T. aestivum × *S. cereale* (4x)	+		85
T. aestivum × *S. cereale* (4x)		0	86
T. aestivum × *S. cereale* (4x)		0	87
T. aestivum × *S. cereale* (4x)	+		88
T. aestivum × *S. cereale* (4x)	+		89
T. aestivum × *S. cereale* (4x)		0	90
T. aestivum × *S. cereale* (4x)	+ / −[a]		91
T. aestivum × *S. cereale* (4x)		−	92
T. aestivum × *S. cereale* (4x)		+	93
T. aestivum × *S. montanum* (4x)	+		85
T. aestivum × *T. speltoides* (4x)	+		83
T. aestivum × *T. speltoides* (4x,8x)	+ / −[a]		94
T. aestivum × *T. timopheevi-Ae. squarrosa* (6x)	−		95
T. durum × *S. cereale* (3x,4x)	+		75
T. turgidum var. durum × *Ae. kotschyi* (4x)	+		96
T. turgidum var. durum × *Ae. cylindrica* (4x)	+		96

[a] Indicates a negative and a positive effect, depending on the genotype of the hybrid used.

Further studies showed that the effect of B chromosomes could, in fact, be modified by the background genotype.[69,72,74] In other words, genotypes of *L. perenne* and *L. rigidum* were discovered which could impose their own constraints upon homoeologous association, even in the absence of B chromosomes.[70] In the amphidiploid, the genes promote homologous bivalent formation and act in a way similar to the Ph locus of wheat. The identification of a similar diploidizing genotype of *L. multiflorum*[66] is seen as the only means of generating stable agronomically desirable amphidiploids with *L. perenne*,[64] in view of the ineffectiveness and non-Mendelian inheritance of B chromosomes in this material.

The effects of these genes and B chromosomes are additive. So it is possible to select contrasting diploid hybrid genotypes both with and without these factors which show very low and very high chiasma frequencies, respectively, at metaphase I (Figure 11). At the tetraploid level, hybrids without these factors will form multivalents at metaphase I, whereas those with them will be effectively diploidized and will behave like allotetraploids, forming largely homologous bivalents (Figure 11). How the effect of these factors is mediated in hybrids between *L. temulentum* and *L. perenne* is the subject of the following section.

B. Genotypic Effect at Zygotene and Pachytene

The suppression of homoeologous association at metaphase I by genes on the normal A chromosomes and accessory B chromosomes can be effected in two ways. Either these factors inhibit synapsis between homoeologs and thus prevent crossing over and chiasma formation, or they inhibit the recombination mechanism itself subsequent to full pachytene pairing. These possibilities were investigated in hybrids between *L. temulentum* and *L. perenne* using the technique of three-dimensional reconstruction of SCs.

If the diploidizing genes and B chromosomes (referred to collectively as diploidizing

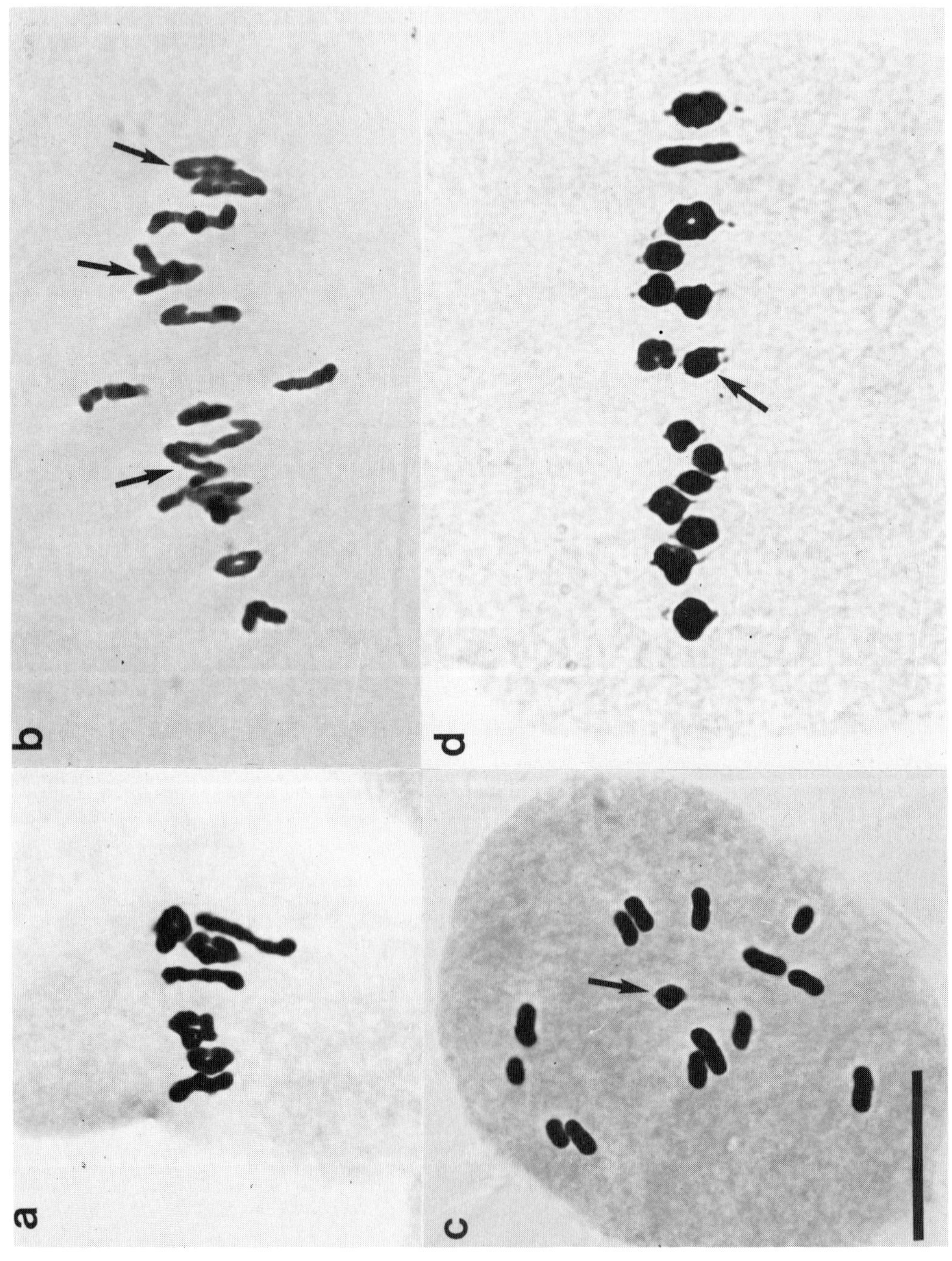

FIGURE 11. Metaphase I in hybrids of *Lolium temulentum* × *L. perenne*. The diploid without diploidizing factors (a) shows complete bivalent formation and a high chiasma frequency. Its tetraploid counterpart (b) has several multivalents (arrows). In contrast, chiasmate association between homoeologs is suppressed in the presence of diploidizing factors, leading to high univalency in the diploid (c) and the promotion of homologous, bivalent formation in the tetraploid (d). Arrows in (c) and (d) indicate the B chromosome bivalent and quadrivalent, respectively. Bar represents 10 μm. (Courtesy of G. M. Evans and M. J. Scanlon.)

factors) are introduced into the high chiasma frequency diploid hybrid described above, chiasmate association at metaphase is, as expected, virtually absent. In fact, Jenkins and Scanlon[26] recorded mean chiasma and bivalent frequencies of only 1.6 and 1.5 per pollen mother cell, respectively, in the plant used in the following analysis. However, the behavior of the chromosomes at metaphase does not accurately reflect the pairing of chromosomes during meiotic prophase. At pachytene, extensive but incomplete SCs are formed between not only homoeologous but also nonhomologous chromosome segments, resulting in the formation of bivalents, multivalents, and foldback loops.[26] Hence, diploidizing factors do not act by completely inhibiting the synapsis of homoeologous chromosomes. Since these configurations do not appear later in meiosis, it follows that in the presence of diploidizing factors, homoeologous as well as nonhomologous SCs are eliminated before metaphase I. The exact timing of much of the elimination was determined by the reconstruction of diplotene nuclei, in which it was shown that the chromosomes were largely unpaired, as at metaphase I. Since many of the illegitimate pairing configurations persist to late pachytene in this hybrid, it was concluded that much of the elimination process occurred upon entry into diplotene, when SCs are subject to normal disintegration processes. Pairing association would be consolidated under normal circumstances by chiasma formation, which would ensure regular disjunction at anaphase I.[97] Clearly, in this hybrid chiasma formation is largely suppressed. It was concluded, therefore, that most of the SCs formed during pachytene between both homoeologous and nonhomologous chromosome segments are not effective in terms of chiasma formation. Diploidizing factors seem to suppress chiasma formation between homoeologous regions and nonhomologous SCs appear to be inherently ineffective, as was seen in the other diploid hybrid. The evidence presented by this hybrid indicates that the diploidizing factors may exert their effect in one of two ways. Even at pachytene, pairing in this hybrid is incomplete, suggesting that the diploidizing factors may be responsible for a localized disruption of SC assembly. Indeed, if this occurs preferentially in distal regions of the chromosomes where crossovers are expected to occur, it would seem likely that chiasma frequencies would be reduced. A correlation between extensive asynapsis and a 50% reduction in chiasma frequencies has been reported in a wheat line triisosomic for the long arm of chromosome 5B, containing six doses of the Ph locus.[98] However, in the *Lolium* hybrid localized asynapsis cannot be the only explanation because long telomeric stretches of SC are often observed, yet do not form chiasmata. Indeed, a preliminary study of surface-spread nuclei, which allows a greater number of nuclei to be studied, indicates that many bivalents form complete SCs from telomere to telomere in hybrids which have very low chiasma frequencies.[99] Alternatively, diploidizing factors may actually suppress crossing over itself within homoeologously paired chromosome segments. In other words, homoeologous SC is, in some way, rendered ineffective by these factors. Such an explanation was offered in a recent study of SC formation in hybrids between *Triticum aestivum* and *T. kotschyi*.[37] At metaphase I in the hybrid containing a mutant allele of Ph, chiasmate association was permitted between homoeologous chromosomes. This contrasts with the normal suppression of such configurations in the presence of the wild-type allele. However, at pachytene there was at least 90% pairing between chromosomes in both genotypes. Clearly, the pH locus operates not by suppressing pairing, but probably by influencing crossing over. In another recent reconstruction analysis of triploid and tetraploid wheat-rye hybrids, a similar conclusion was reached[36] because low chiasma frequencies at metaphase I appeared unrelated to the extensive SC formation during meiotic prophase.

Support for the notion that the Ph-like factors of *Lolium* probably work by inhibiting crossing over has come from further studies of the amphiploids. In the colchicine-doubled amphidiploid without diploidizing factors, extensive chiasmate association is permitted between homoeologs at metaphase I in the form of trivalents and quadrivalents. Similar configurations were observed at late zygotene, although at this stage nonhomologous SC joined chromosomes into multivalents with valences higher than four.[28] As in its diploid hybrid counterpart, nonhomologous SC does not support crossing over and is eliminated before metaphase I. If B chromosomes and diploidizing genes are introduced, there is, as expected, a dramatic decrease in the frequency of multivalents at metaphase I and a commensurate increase in the number of homologous bivalents. This effect is not achieved by a restriction of synapsis to homologous chromosomes only, for at early zygotene and pachytene both homeologous and nonhomologous chromosomes pair in multiple configurations. Since these do not appear at metaphase I, the most likely explanation is that the diploidizing factors are responsible for the elimination of homoeologous SC, probably by suppressing crossing over. This is not to say that this is the only means of ensuring exclusive bivalent formation because there is evidence that multivalents may be resolved to bivalents from the beginning of pachytene. The elimination of multivalents by strict control of crossing over is the mechanism of bivalent formation in normal allohexaploid wheat.[100-102] It is remarkable how closely the effects of the diploidizing factors mirror those of the Ph locus from both the genetical and cytological point of view.

Study of a triploid hybrid containing two sets of *L. perenne* chromosomes, one set of *L. temulentum* chromosomes, and two B chromosomes revealed another interesting aspect of chromosome pairing behavior in this material.[27] Under the influence of the B chromosomes, chiasmate association at metaphase I is restricted entirely to the homologous *L. perenne* chromosomes, the *L. temulentum* set being present as univalents. At zygotene, as may be expected, chromosome synapsis does not suffer the same restriction and homologs are permitted association with homoeologs with, in this case, little nonhomologous SC formation. At a later stage, multivalents are transformed to bivalents and the *L. temulentum* chromosomes synapse extensively both within and between themselves. It is as if there were a relaxation in the requirement for homology and the univalents fulfil their propensity for pairing by forming ineffective SCs. The division of meiotic prophase into homologous and nonhomologous pairing phases was also observed in achiasmate autotriploid *Bombyx* oocytes.[103] The disappearance of duplication[48] and inversion[104] loops by late pachytene in mice may be another manifestation of the same process.

The elimination of homoeologous SC under the influence of the diploidizing factors and the inherent ineffectiveness of nonhomologous SC can explain the origin of all the chromosome configurations at metaphase I in the *Lolium* hybrids.[105] However, the preceding studies fail to explain how the factors actually influence the siting of crossovers. It is also unknown how some chromosomes of the complement can form quite legitimate associations right from the beginning of synapsis without the assistance or intervention of any mechanism which monitors and modifies chromosome pairing behavior and chiasma formation. To understand how much nuclear processes operate would undoubtedly require a detailed analysis of the biochemistry of chromosome pairing and recombination.

IV. BIOCHEMISTRY OF CHROMOSOME PAIRING

An account of chromosome pairing and fertility in hybrids would not be complete without mention of some of the pioneering work on the biochemistry of meiosis and, more specifically, on the biochemistry of chromosome pairing itself. Much of the work was performed on lily hybrids. Following the discovery of DNA synthesis during meiosis,[106] Ito et al.[107] tested its functional significance in a diploid hybrid between *Lilium aureliensis* and *L. henryi* cv. Cinnabar. Several inhibitors of DNA synthesis were administered to cultured meiocytes of the hybrid during zygotene and pachytene. These caused arrest of zygotene or fragmentation of chromosomes, depending on their precise time of application, observations that led to the conclusion that there was a correlation between cytological effects and the synthesis of DNA during this period. Studies in the same hybrid using the inhibitor of DNA synthesis, deoxyadenosine, indicated that DNA synthesis during the synaptic interval is required to initiate SC formation and is necessary for normal association at diplotene and regular separation of sister chromatids at anaphase II.[108] Similarly, if DNA synthesis is inhibited by mitomycin c during leptotene and zygotene, the formation of the SC is prevented.[109] Protein synthesis, too, is required for chromosome pairing and chiasma formation. If cells are exposed to cycloheximide during late zygotene or early pachytene, the SC is irreversibly disrupted[110] and chiasmata are prevented from forming.[111] If the protein sythesis inhibitor is applied slightly earlier and the cells then cultured in a cycloheximide-free medium, normal SCs may reform,[110] but the chromatin may be reorganized in a different way.[109] Evidently, both DNA and protein synthesis are integral parts of the process of chromosome pairing and recombination.

Colchicine, a spindle inhibitor, has also been shown to have an effect on synapsis and chiasma formation in microsporocytes of the triploid hybrid cultivar, *L. aureliensis* × *L. henryi*. If it is applied up to early zygotene, partial asynapsis is induced, with a reduction in chiasma frequency and an increase in the number of univalents at metaphase I.[112] A significant reduction in chiasma frequency at metaphase I under the influence of colchicine was noted even in a normally largely achiasmate diploid hybrid between the same two species.[113] However, this effect was not accompanied by a disruption in the relative amounts of SC in treated and untreated cells. Furthermore, colchicine-doubled tetraploid cells displayed high chiasma frequencies and almost complete bivalent formation. This shows that the colchicine effect does not persist and disrupt chiasma formation between homologous chromosomes. It also shows that the achiasmate condition of the diploid hybrid was attributable not to some genic defect, but to insufficient homology for effective pairing. The latter conclusion had important implications for an unprecedented investigation into the link between the actual process of homologous pairing and the regulation of specific metabolic activities.

In chiasmate hybrids of lily, it has been shown that the activities of certain meiosis-specific proteins undergo cyclic change during meiotic prophase.[114] Since their activities rise at about the beginning of leptotene, reach a peak during zygotene and pachytene, and fall off rapidly towards the end of prophase, it has been presumed that they play a direct role in chromosome pairing and recombination.[115]

In the achiasmate lily hybrid described above, the behavior of two of these enzymes is different.[116] One of the proteins is an endonuclease responsible for nicking the DNA during pachytene as a prelude to genetic recombination.[117] In the achiasmate hybrid, no such nicking occurs despite appropriate enzyme activity. Similarly, the activity of a protein which facilities DNA reannealing[118,119] is severely reduced in the same hybrid. However, when the chromosome number is doubled and homologous pairing and chiasma formation occurs, the normal activities of these two enzymes are restored.[116] The conclusions were that effective homologous pairing per se was the only condition required for restoring normal meiotic metabolism in achiasmate meiocytes of the hybrid.

V. CONCLUDING REMARKS

In addition to their widespread use in breeding programs, hybrids have provided and continue to provide the cytologist with a tool for unravelling the mysteries of the meiotic process. The disruption of chromosome pairing due to hybridity and its effect on chiasma frequency and fertility has shed much light on the nature of synapsis and recombination in nonhybrid plants too and has elucidated some of the constraints upon the evolution of the chromosomes themselves. Homoeologous chromosomes of hybrids can show a remarkable capacity to pair from end to end and to form chiasmata, despite considerable structural divergence. Sufficient homology is preserved within these bivalents to permit genetic exchange, but the homoeologous DNA molecules present large enough differences for Ph-like factors to discriminate between strictly homologous and homoeologous paired regions and to control accordingly recombination between particular chromosomes. In view of the greater understanding of the biochemistry of meiosis, hopefully such investigations will prompt research into the exact molecular function of these diploidizing factors. This could have important implications for the development and exploitation of new and reliable hybrid plant species.

ACKNOWLEDGMENTS

I thank Professor H. Rees for his help in the preparation of this manuscript and I am grateful to R. N. Jones, S. M. Albini, A. G. Seal, G. M. Evans, and M. J. Scanlon for providing material for publication.

REFERENCES

1. **Michaelis, P.,** Cytoplasmic inheritance in *Epilobium* and its thoeretical significance, *Adv. Genet.,* 6, 288, 1954.
2. **Rees, H. and Jones, R. N.,** Structural basis of quantitative variation in nuclear DNA, *Nature,* 216, 825, 1967.
3. **Jones, R. N. and Rees, H.,** Nuclear DNA variation in *Allium, Heredity,* 23, 591, 1968.
4. **Albini, S. M.,** Synaptonemal Complex Studies in *Allium* Species, Ph.D. thesis, University of Birmingham, Birmingham, England, 1986.
5. **Ernst, H.,** Zytogenetische Untersuchungen an *Antirrhinum majus* L. *Z. Bot.,* 34, 81, 1939.
6. **Bharathi, M., Murty, U. R., Kirti, P. B., and Rao, N. G. P.,** Chromosome pairing in the interspecific hybrids: *Arachis hypogaea* L. × *Arachis chacoense* nom. nud. Krap. et Greg. and *A. hypogaea* L. × *A. villosa* Benth., *Cytologia,* 48, 527, 1983.
7. **Kirty, P. B., Murty, U. R., Bharathi, M., and Rao, N. G. P.,** Chromosome pairing in F1 hybrid *Arachis hypogaea* L. × *A. monticola* Krap. et Rig., *Theor. Appl. Genet.,* 62, 139, 1982.
8. **Tobgy, H. A.,** A cytological study of *Crepis fuliginosa, C. neglecta,* and their F1 hybrid, and its bearing on the mechanism of phylogenetic reduction in chromosome number, *J. Genet.,* 45, 67, 1943.
9. **Israel, H. W. and Sagawa, Y.,** Post-pollination ovule development in *Dendrobium* orchids. III. Fine structure of meiotic prophase I, *Caryologia,* 18, 15, 1965.
10. **Longley, A. E.,** Morphological characters of teosinte chromosomes, *J. Agri. Res.,* 54, 835, 1937.
11. **Seal, A. G. and Rees, H.,** The distribution of quantitative DNA changes associated with the evolution of diploid *Festuceae, Heredity,* 49, 179, 1982.
12. **Rees, H., Jenkins, G., and Seal, A. G.,** Quantitative DNA variation and chromosome homology, in *Controlling Events in Meiosis,* Evans, C. W. and Dickinson, H. G., Eds., Company of Biologists, Cambridge, 1984, 321.
13. **Seal, A. G.,** Cytogenetic Studies in *Festuca and Lolium,* Ph.D. thesis, University College of Wales, Aberystwyth, Wales, 1981.
14. **Seal, A. G.,** The distribution and consequences of changes in nuclear DNA content, in *Kew Chromosome Conference II,* Brandham, P. E. and Bennett, M. D., Eds., Allen and Unwin, London, 1983, 225.
15. **Jenkins, G.,** The Consequences of Interspecific DNA Variation Upon Chromosome Behaviour, Ph.D. thesis, University College of Wales, Aberystwyth, Wales, 1981.
16. **Jenkins, G. and Rees, H.,** Synaptonemal complex formation in a *Festuca* hybrid, in *Kew Chromosome Conference II,* Brandham, P. E. and Bennett, M. D., Eds., Allen and Unwin, London, 1983, 233.
17. **Rees, H., Jenkins, G., Seal, A. G., and Hutchinson, J.,** Assays of the phenotypic effects of changes in DNA amounts, in *Genome Evolution,* Academic Press, London, 1982, 287.
18. **Brown, M. S.,** A comparison of pachytene and metaphase pairing in species hybrids of *Gossypium, Genetics,* 39, 962, 1954.
19. **Beasley, J. D. and Brown, M. S.,** Asynaptic plants and their polyploids, *J. Agric. Res.,* 65, 421, 1942.
20. **Sen, S. K.,** Axial elements in plant meiotic chromosomes, *Nature,* 228, 79, 1970.
21. **La Cour, L. F. and Wells, B.,** Abnormalities in synaptonemal complexes in pollen mother cells of a lily hybrid, *Chromosoma,* 42, 137, 1973.

22. **La Cour, L. F. and Wells, B.,** The chromomeres of prepachytene chromosomes, *Cytologia*, 36, 111, 1971.
23. **Rees, H., Cameron, F. M., Hazarika, M. H., and Jones, G. H.,** Nuclear variation between diploid angiosperms, *Nature*, 211, 828, 1966.
24. **Rees, H. and Jones, G. H.,** Chromosome evolution in *Lolium*, *Heredity*, 22, 1, 1967.
25. **Jenkins, G.,** Synaptonemal complex formation in hybrids of *Lolium temulentum* × *Lolium perenne (L.).* I. High chiasma frequency diploid, *Chromosoma*, 92, 81, 1985.
26. **Jenkins, G. and Scanlon, M. J.,** Chromosome pairing in a *Lolium temulentum* × *Lolium perenne* diploid hybrid with a low chiasma frequency, *Theor. Appl. Genet.*, 73, 516, 1987.
27. **Jenkins, G.,** Synaptonemal complex formation in hybrids of *Lolium temulentum* × *Lolium perenne (L.).* II. Triploid, *Chromosoma*, 92, 387, 1985.
28. **Jenkins, G.,** Synaptonemal complex formation in hybrids of *Lolium temulentum* × *Lolium perenne (L.).* III. Tetraploid, *Chromosoma*, 93, 413, 1986.
29. **Jenkins, G.,** Chromosome pairing in *Lolium* hybrids, in *Kew Chromosome Conference III*, in press.
30. **Menzel, M. Y.,** Pachytene chromosomes of the intergeneric hybrid *Lycopersicon esculentum* × *Solanum lycopersicoides*, *Am. J. Bot.*, 49, 605, 1962.
31. **Menzel, M. Y. and Price, J. M.,** Fine structure of synapsed chromosomes in F1 *Lycopersicon esculentum — Solanum lycopersicoides* and its parents, *Am. J. Bot.*, 53, 1079, 1966.
32. **Shastry, S. V. S., Smith, W. K., and Cooper, D. C.,** Chromosome differentiation in several species of *Melilotus*, *Am. J. Bot.*, 47, 613, 1960.
33. **Gottschalk, W. and Peters, N.,** Das Konjugationsverhalten partiell homologer Chromosomen, *Chromosoma*, 7, 708, 1956.
34. **Von Wangenheim, K. H.,** Das Pachytaen und der weitere Ablauf der Meiose in diploiden *Solanum* Arten und Bastarden, *Chromosoma*, 8, 671, 1957.
35. **Bennett, M. D., Smith, J. B., Simpson, S., and Wells, B.,** Intranuclear fibrillar material in cereal pollen mother cells, *Chromosoma*, 71, 289, 1979.
36. **Abirached-Darmency, M., Cauderon, Y., and Zickler, D.,** Meiotic chromosome pairing in three F1: *(Triticum-Secale)* hybrids. A comparative approach in light and electron microscopy, *Biol. Cell.*, 51, 365, 1984.
37. **Gillies, C. B.,** The effect of Ph gene alleles on synaptonemal complex formation in *Triticum aestivum* × *T. kotschyi* hybrids, *Theor. Appl. Genet.*, 74, 430, 1987.
38. **Maguire, M. P.,** A study of homology between a terminal portion of *Zea* chromosome 2 and a segment derived from *Tripsacum*, *Genetics*, 45, 195, 1960.
39. **Maguire, M. P.,** A study of pachytene chromosome pairing in a corn-*Tripsacum* hybrid derivative, *Genetics*, 45, 651, 1960.
40. **Maeda, T.,** Chiasma studies in *Allium fistolosum, Allium cepa* and their F1, F2 and backcross hybrids, *Jpn. J. Genet.*, 13, 146, 1937.
41. **Maeda, T.,** Chiasma studies in *Allium*, *Jpn. J. Genet.*, 12, 163, 1942.
42. **Rees, H.,** Nuclear DNA variation and the homology of chromosomes, in *Plant Biosystematics*, Academic Press, Toronto, 1984, 87.
43. **Counce, S. J. and Meyer, G. F.,** Differentiation of the synaptonemal complex and kinetochore in *Locusta* spermatocytes studies by whole mount electron microscopy, *Chromosoma*, 44, 231, 1973.
44. **Albini, S. M., Jones, G. H., and Wallace, B. M. N.,** A method for preparing two-dimensional surface spreads of synaptonemal complexes from plant meiocytes for light and electron microscopy, *Exp. Cell Res.*, 152, 280, 1984.
45. **Albini, S. M. and Jones, G. H.,** Synaptonemal complex-associated centromeres and recombination nodules in plant meiocytes prepared by an improved surface-spreading technique, *Exp. Cell Res.*, 155, 588, 1984.
46. **Seal, A. G.,** DNA variation in *Festuca*, *Heredity*, 50, 225, 1983.
47. **Jenkins, G. and Bennett, M. D.,** The intranuclear relationship between centromere volume and chromosome size in *Festuca scariosa* × *drymeja*, *J. Cell Sci.*, 47, 117, 1981.
48. **Moses, M. J. and Poorman, P. A.,** Synaptonemal complex analysis of mouse chromosome rearrangements. II. Synaptic adjustment in a tandem duplication, *Chromosoma*, 81, 519, 1981.
49. **Riley, R. and Flavell, R. B.,** A first view of the meiotic process, *Philos. Trans. R. Soc. London Ser. B*, 277, 191, 1977.
50. **Rees, H. and Dale, P. J.,** Chiasmata and variability in *Lolium* and *Festuca* populations, *Chromosoma*, 47, 335, 1974.
51. **Hutchinson, J., Rees, H., and Seal, A. G.,** An assay of the activity of supplementary DNA in *Lolium*, *Heredity*, 43, 411, 1979.
52. **Okamoto, M.,** Asynaptic effect of chromosome V, *Wheat Inf. Serv.*, 5, 6, 1957.
53. **Riley, R. and Chapman, V.,** Genetic control of the cytologically diploid behaviour of hexaploid wheat, *Nature*, 182, 713, 1958.

54. **Jones, R. N. and Rees, H.**, *B-chromosomes*, Academic Press, London, 1982.
55. **Mochizuki, A.**, Further studies on the effect of accessory chromosomes on chromosome pairing in *Aegilops mutica*, *Jpn. J. Genet.*, 39, 356, 1964.
56. **Ohta, S. and Tanaka, M.**, Reconsideration of the genome of *Aegilops mutica* Boiss. based on the chromosome pairing in interspecific and intergeneric hybrids, *Wheat Inf. Serv.*, 52, 33, 1981.
57. **Sano, J. and Tanaka, T.**, Estimation of chromosomal homology between *Aegilops speltoides* and the tetraploid wheats by using B chromosomes, *Jpn. J. Genet.*, 55, 9, 1980.
58. **Rajhathy, T. and Thomas, H.**, Genetic control of chromosome pairing in hexaploid oats, *Nature New Biol.*, 239, 217, 1972.
59. **Evans, G. M. and Aung, T.**, The influence of the genotype of *Lolium perenne* on homoeologous chromosome association in hexaploid *Festuca arundinacea*, *Heredity*, 56, 97, 1986.
60. **Jauhar, P. P.**, Chromosome relationship between *Lolium* and *Festuca (Gramineae)*, *Chromosoma*, 52, 103, 1975.
61. **Jauhar, P. P.**, Chromosome pairing in some triploid and trispecific hybrids in *Lolium-Festuca* and its phylogenetic implications, in *Chromosomes Today*, Vol. 5, Pearson, P. L. and Lewis, K. R., Eds., John Wiley & Sons, Jerusalem, 1976, 165.
62. **Hovin, A. W. and Hill, D. B.**, B chromosomes, their origin and relation to meiosis in interspecific *Lolium* hybrids, *Am. J. Bot.*, 53, 702, 1966.
63. **Evans, G. M. and Macefield, A. J.**, The effect of B chromosomes on homoeologous pairing in species hybrids. II. *Lolium multiflorum* × *Lolium perenne*, *Chromosoma*, 45, 369, 1974.
64. **Aung, T. and Evans, G. M.**, The potential for diploidising *Lolium multiflorum* × *L. perenne* tetraploids, *Can. J. Genet. Cytol.*, 27, 506, 1985.
65. **Bowman, J. G. and Thomas, H.**, B-chromosomes and chromosome pairing in *Lolium perenne* × *Festuca arundinacea* hybrid, *Nature New Biol.*, 245, 80, 1973.
66. **Evans, G. M. and Aung, T.**, Identification of a diploidising genotype of *Lolium multiflorum*, *Can. J. Genet. Cytol.*, 27, 498, 1985.
67. **Evans, G. M. and Macefield, A. J.**, Supression of homoeologous pairing by B chromosomes in a *Lolium* species hybrid, *Nature New Biol.*, 236, 110, 1972.
68. **Evans, G. M. and Macefield, A. J.**, The effect of B chromosomes on homoeologous pairing in species hybrids. I. *Lolium temulentum* × *Lolium perenne*, *Chromosoma*, 41, 63, 1973.
69. **Evans, G. M. and Taylor, I. B.**, Genetic control of homoeologous chromosome pairing in *Lolium* hybrids, in *Current Chromosome Research*, Jones, K. and Brandham, P. E., Eds., Elsevier/North Holland, Amsterdam, 1976, 57.
70. **Taylor, I. B. and Evans, G. M.**, The genetic control of homoeologous chromosome association in *Lolium temulentum* × *Lolium perenne* interspecific hybrids, *Chromosoma*, 62, 57, 1977.
71. **Evans, G. M. and Macefield, A. J.**, Induced translocation of an A and a B chromosome in *Lolium perenne*, *Chromosoma*, 61, 257, 1977.
72. **Evans, G. M. and Davies, E. W.**, The genetics of meiotic chromosome pairing in *Lolium temulentum* × *Lolium perenne* tetraploids, *Theor. Appl. Genet.*, 71, 185, 1985.
73. **Aung, T. and Evans, G. M.**, Segregation of isoenzyme markers and meiotic pairing control genes in *Lolium*, *Heredity*, 59, 129, 1987.
74. **Taylor, I. B. and Evans, G. M.**, The effect of B chromosomes on homeologous pairing in species hybrids. III. Intraspecific variation, *Chromosoma*, 57, 25, 1976.
75. **Naranjo, T., Lacadena, J. R., and Giraldez, R.**, Interaction between wheat and rye genomes on homologous and homoeologous pairing, *Z. Pflanzenzuecht.*, 82, 289, 1979.
76. **Jouve, N. and Giorgi, B.**, Analysis of induced homoeologous pairing in hybrids between 6x triticale ph 1 mutant and *Triticum aestivum* L., *Can. J. Genet. Cytol.*, 28, 696, 1986.
77. **Mello-Sampayo, T.**, Promotion of homoeologous pairing in hybrids of *Triticum aestivum* × *Aegilops longissima*, *Genet. Iber.*, 23, 1, 1971.
78. **Riley, R., Kimber, G., and Chapman, V.**, The origin of the genetic control of diploid like behavior of polyploid wheat, *J. Hered.*, 52, 22, 1961.
79. **Dover, G. A. and Riley, R.**, Prevention of pairing of homoeologous meiotic chromosomes of wheat by an activity of supernumerary chromosomes of *Aegilops*, *Nature*, 240, 159, 1972.
80. **Dover, G. A. and Riley, R.**, Variation at two loci affecting homoeologous meiotic pairing in *Triticum aestivum* × *Aegilops mutica* hybrids, *Nature New Biol.*, 235, 61, 1972.
81. **Vardi, A. and Dover, G. A.**, The effect of B chromosomes on meiotic and pre-meiotic spindles and chromosome pairing in *Triticum/Aegilops* hybrids, *Chromosoma*, 38, 367, 1972.
82. **Dvorak, J.** Genetic variability in *Aegilops speltoides* affecting homoeologous pairing in wheat, *Can. J. Genet. Cytol.*, 14, 371, 1972.
83. **Feldman, M.**, The regulation of chromosome pairing in inter-varietal, inter-specific and inter-generic hybrids of common wheat, *EWAC Newslett.*, 3, 2, 1971.

84. **Wall, A. M., Riley, R., and Chapman, V.,** Wheat mutants permitting homoeologous chromosome pairing, *Genet. Res.,* 18, 311, 1971.

85. **Lelley, T.,** Induction of homoeologous pairing in wheat by genes of rye suppressing chromosome 5B effect *Can. J. Genet. Cytol.* 18, 485, 1976.

86. **Lelley, T.,** The effect of supernumerary chromosomes of rye on homeologous pairing in hexaploid wheat, *Z. Pflanzenzuecht.,* 77, 281, 1976.

87. **Roothan, M. and Sybenga, J.,** No 5B compensation by C-chromosomes, *Theor. Appl. Genet.,* 48, 63, 1976.

88. **Dvorak, J.,** Effect of rye on homoeologous chromosome pairing in wheat × rye hybrids, *Can. J. Genet. Cytol.,* 19, 549, 1977.

89. **Dhaliwal, H. S., Gill, B. S., and Waines, J. G.,** Analysis of induced homoeologous pairing in a ph mutant wheat × rye hybrid, *J. Hered.,* 68, 206, 1977.

90. **Nejzing, M. G. and Viegas, W. S.,** The effect of rye B chromosomes on meiotic stability of rye-wheat hybrids in normal, nulli-5B and nulli-5D background, *Genetica,* 51, 21, 1979.

91. **Romero, C. and Lacadena, J. R.,** Interaction between rye B-chromosomes and wheat genetic systems controlling homoeologous pairing, *Chromosoma,* 80, 33, 1980.

92. **Viegas, W. S.,** The effect of B-chromosome of rye on the chromosome association in F1 hybrids *Triticum aestivum × Secale cereale* in the absence of chromosome 5B or 5D, *Theor. Appl. Genet.,* 56, 193, 1980.

93. **Romero, C. and Lacadena, J. R.,** Effect of rye B chromosomes on pairing in *Triticum aestivum × Secale cereale* hybrids, *Z. Pflanzenzeucht.,* 89, 39, 1982.

94. **Kimber, G. and Athwal, R. S.,** A reassessment of the course of evolution of wheat, *Proc. Natl. Acad. Sci. U.S.A.,* 69, 912, 1972.

95. **Feldman, M.,** The mechanism regulating pairing in *Triticum timopheevi, Wheat Inf. Serv.,* 21, 1,

96. **Giorgi, B. and Barbera, F.,** Increase of homoeologous pairing in hybrids between a ph mutant of *Triticum turgidum L. var durum* and two tetraploid species of *Aegilops: Aegilops kotschyi* and *Ae. cylindrica, Cereal Res. Commun.,* 9, 205, 1981.

97. **Rasmussen, S. W. and Holm, P. B.,** Chromosome pairing in autotetraploid *Bombyx* females. Mechanism for exclusive bivalent formation, *Carlsberg Res. Commun.,* 44, 101, 1979.

98. **Wischmann, B.,** Chromosome pairing and chiasma formation in wheat plants triisosomic for the long arm of chromosome 5B, *Carlsberg Res. Commun.,* 51, 1, 1986.

99. **Jenkins, G.,** unpublished observations.

100. **Hobolth, P.,** Chromosome pairing in allohexaploid wheat var. Chinese Spring. Transformation of multivalents into bivalents, a mechanism for exclusive bivalent formation, *Carlsberg Res. Commun.,* 46, 129, 1981.

101. **Jenkins, G.,** Chromosome pairing in *Triticum aestivum* cv. Chinese Spring, *Carlsberg Res. Commun.,* 48, 255, 1983.

102. **Holm, P. B.,** Chromosome pairing and chiasma formation in allohexaploid wheat, *Triticum aestivum* analyzed by spreading of meiotic nuclei, *Carlsberg Res. Commun.,* 51, 239, 1986.

103. **Rasmussen, S. W.,** Chromosome pairing in triploid females of *Bombyx mori* analyzed by three dimensional reconstructions of synaptonemal complexes, *Carlsberg Res. Commun.,* 42, 163, 1977.

104. **Moses, M. J., Poorman, P. A., Roderick, T. H., and Davisson, M. T.,** Synaptonemal complex analysis of mouse chromosomal rearrangements. IV. Synapsis and synaptic adjustment in two paracentric inversions, *Chromosoma,* 84, 457, 1982.

105. **Jenkins, G.,** Chromosome pairing in *Lolium* hybrids, in *Kew Chromosome Conference III,* in press.

106. **Hotta, Y., Ito, M., and Stern, H.,** Synthesis of DNA during meiosis, *Proc. Natl. Acad. Sci. U.S.A.,* 56, 1184, 1966.

107. **Ito, M., Hotta, Y., and Stern, H.,** Studies of meiosis in vitro. II. Effect of inhibiting DNA synthesis during meiotic prophase on chromosome structure and behavior, *Dev. Biol.,* 16, 54, 1967.

108. **Roth, T. F. and Ito, M.,** DNA-dependent formation of the synaptonemal complex at meiotic prophase, *J. Cell Biol.,* 35, 247, 1967.

109. **Sen, S. K.,** Chromosome organization during and after synapsis in cultured microsporocytes of *Lilium* in presence of mitomycin C and cycloheximide, *Exp. Cell Res.,* 55, 123, 1969.

110. **Roth, T. F. and Parchman, L. G.,** Alteration of meiotic chromosomal pairing and synaptonemal complexes by cycloheximide, *Chromosoma,* 35, 9, 1971.

111. **Parchman, L. G. and Stern, H.,** The inhibition of protein synthesis in meiotic cells and its effect on chromosome behavior, *Chromosoma,* 26, 298, 1969.

112. **Shepard, J., Boothroyd, E. R., and Stern, H.,** The effect of colchicine on synapsis and chiasma formation in microsporocytes of *Lilium, Chromosoma,* 44, 423, 1974.

113. **Toledo, L. A., Bennett, M. D., and Stern, H.,** Cytological investigations of the effect of colchicine on meiosis in *Lilium* hybrid cv. 'Black Beauty' microsporocytes, *Chromosoma,* 72, 157, 1979.

114. **Stern, H. and Hotta, Y.,** Biochemistry of meiosis, *Philos. Trans. R. Soc. London Ser. B,* 277, 277, 1977.

115. **Stern, H. and Hotta, Y.,** Regulatory mechanisms in meiotic crossing-over, *Annu. Rev. Plant Physiol.,* 29, 415, 1978.
116. **Hotta, Y., Bennett, M. D., Toledo, L. A., and Stern, H.,** Regulation of R-protein and endonuclease activities in meiocytes by homologous chromosome pairing, *Chromosoma,* 72, 191, 1979.
117. **Howell, S. H. and Stern, H.,** The appearance of DNA breakage and repair activities in the synchronous meiotic cycle of *Lilium, J. Mol. Biol.,* 55, 357, 1971.
118. **Hotta, Y. and Stern, H.,** A DNA-binding protein in meiotic cells of *Lilium, Dev. Biol.,* 26, 87, 1971.
119. **Mather, J. and Hotta, Y.,** A phosphorylatable DNA-binding protein associated with a lipoprotein fraction from rat spermatocyte nuclei, *Exp. Cell Res.,* 109, 181, 1977.

Chapter 5

CHROMOSOME PAIRING AND FERTILITY IN POLYPLOIDS

C. B. Gillies

TABLE OF CONTENTS

I. INTRODUCTION

Polyploidy is the presence in cells or tissues of an organism of three or more copies of the basic set of chromosomes (the haploid genome). Occasional polyploid cells occur in many types of tissues in both plants and animals. Many important crop plants are polyploid, and polyploidy has undoubtedly been important in the evolution of many plants and probably some animals. More than one third of angiosperms are polyploid[1] and 70 to 80% may have polyploidy in their ancestry. In contrast, polyploidy in sexually reproducing animals is rare. The usual explanation advanced is that polyploidy would disturb the sex chromosome balance, leading to intersexes,[2,3] but this is not always the case. Nevertheless, approximately 10% of human zygotes are polyploid, although most spontaneously abort and the rare liveborns do not survive long.[4,5] Sexually reproducing polyploid species do occur naturally among the amphibia[6-8] and fish,[9,10] but polyploidy in animals is often associated with parthenogenic reproduction, e.g., amphibia, fish, reptiles,[11] nematodes,[12,13] earthworms,[14] and insects.[15]

Polyploidy in general leads to an increase in cell size, which may be accompanied by an increase in the size of the whole organism or specific organs. Other cell and chromosome parameters may be subject to changes disproportionate to ploidy level.[16,17] In addition, the duplication of genes or the combining of genes from related species, often allows for increased diversity. Not all polyploids, however, are superior to their diploid ancestors; higher polyploids, in particular, may exhibit abnormalities.[3] One of the problems which is often encountered by polyploids is a reduction in fertility, much of which can be traced to disturbances in chromosome behavior during meiosis, leading to unbalanced and inviable gametes, although irregular pairing itself may have an effect.[18] In general, the segregational behavior of chromosomes at anaphase I is determined by the type of pairing at metaphase I (univalents, bivalents, trivalents, quadrivalents, etc). This paper addresses the mechanisms of meiotic pairing and crossing over in polyploids and how these can affect fertility.

Polyploidy can be subdivided into at least two types. In *autopolyploids*, only one basic genome is present as multiple copies, while in *allopolyploids*, genomes from two or more species origins are present, presumably following a hybridization event. Since the generation of viable hybrids and allopolyploids requires some phylogenetic relationship between the parental species, it is expected that some homology will exist between chromosomes of the genomes in an allopolyploid. This has resulted in the introduction of the term *segmental allopolyploid* to describe a polyploid containing partially homologous (*homoeologous*) genomes.[19] The existence of partially homologous chromosomes in allopolyploids allows for variability in pairing behavior so that in such cases pairing may be between identical homologues from one genome type — autosyndesis, or it may be between partially homologous chromosomes from different genomes — allosyndesis.[20] While in practice it is often difficult to classify many polyploids as strict autopolyploids or allopolyploids, for convenience I deal separately with issues peculiar to each type.

II. TRIPLOIDS

A. Prophase I Pairing

In the majority of organisms, the light-microscopic study of the zygotene and pachytene stages is difficult and, hence, there have been very few reports of extensive studies of these stages in autopolyploids. Light-microscopic data from later meiotic stages have often been used to predict *post hoc* the behavior in the preceding stages (see below). However, the discovery that meiotic pairing during zygotene and pachytene is mediated by the synaptonemal complex, a structure only clearly distinguishable by electron microscopy,[21,22] has resulted more recently in ultrastructural studies of early prophase I pairing of polyploids.

Because synapsis in polyploids can involve (pairwise) associations of three or more chromosomes at different points along their lengths, electron microscopy of random sections reveals little about the process. Such studies of triploid *Rhoeo spathacea*[23] and triploid and tetraploid *Allium amplectans*[24] merely confirmed that such polyploids could form apparently normal synaptonemal complex segments. It required the application of serial section reconstruction and, later, spreads of whole nuclei to reveal the ultrastructural details of autopolyploid pachytene pairing.

Moens was the first to apply serial sectioning to the study of synaptonemal complex formation in polyploids. Following a detailed examination of the process of synaptonemal complex formation in diploid lily,[25] he serially sectioned and reconstructed individual multivalent synaptonemal complexes from triploid *Lilium tigrinum*[26-28] and autotetraploid *L. longiflorum.*[28] In *L. tigrinum*, which may in fact be a segmental allotriploid, trivalent figures were seen in reconstructed zygotene/pachytene nuclei. In one or more sections, two of the lateral elements were associated in a synaptonemal complex, while the third unpaired axial element usually lay close by.[26,27] In subsequent sections switches of pairing partner occurred, so that the unpaired axial element became one laternal element of a synaptonemal complex and one of the former lateral elements was now present as an unpaired axial element. While the partner switches seen in some trivalents were, as expected on the basis of only pairwise axial element associations, forming synaptonemal complexes, in some trivalent synaptonemal complexes there was close association of three axial elements and even apparent fusion of two axial elements over short distances.[27]

The origin of *L. tigrinum* from two or more closely related species presumably results in the ability of all three members of a homoeologous chromosome set to behave in pairing as strict homologs and to form trivalents in many cases. The three axial elements of one homoeologous chromosome set all passed through one centromere region, two as lateral elements of a synaptonemal complex and one as an unpaired axial element. However, Moens[26,28] did detect abnormalities such as thickened, branched, and tubular lateral elements in some synaptonemal complexes. These he interpreted as the result of pairing between axial elements of different species origin having "genetic inequalities".[28] La Cour and Wells[29,30] studied sections of triploid *Phaedranassa virdifolia* microsporocytes and identified deformities of the lateral elements, which they concluded were caused by asynchronous contraction. They also illustrate an apparent lateral element partner switch in a trivalent synaptonemal complex. Subsequent descriptions of similar lateral element abnormalities in diploid species not of hybrid origin (see Loidl[31] for summary) cast some doubt on the above conclusions.

Only a limited number of studies have been carried out on synaptonemal complex formation behavior at the whole nucleus level in triploids. Serial section studies of autotriploid *Bombyx*[32] and *Coprinus*[33] and a spreading study of *A. sphaerocephalon*[34] will be discussed here. A spreading study of triploid chickens by Comings and Okada[35] illustrated trivalent synaptonemal complexes, some of which exhibited "triple pairing", associations of three lateral elements with two central elements. This will be discussed in more detail below, as will allotriploid and allopentaploid hybrids.

Using such criteria as nuclear vacuole presence lateral element bouquet formation, and nuclear envelope attachment, Rasmussen[32] placed in temporal sequence 11 completely reconstructed oocyte nuclei from serially sectioned triploid silkworm (*Bombyx mori*) females. He found that in the earliest prophase I stage examined the three lateral elements of homologs were associated by telomeric attachments approximately 100 to 200 nm from each other on the nuclear envelope. Pairing was 68% complete at this stage. Synaptonemal complex (SC) formation appeared to start simultaneously from both ends of each homolog set and generally involved two of the three homologous lateral elements so that a trivalent with one pairing partner switch could form. Two nuclei at this stage had 4 and 8 trivalents, respectively, out of a potential maximum of 28 trivalents. In the remainder of cases, two homologous lateral elements had paired to form a bivalent SC, and the third homologous axial element was unpaired, but roughly aligned with the bivalent along most of its length. Occasional examples of triple pairing were noted over short distances near telomeric nuclear envelope attachments.[32] One or both of the central elements of the SC thus formed often lacked the ladder-like structure characteristic of SCs with two normally paired lateral elements. The frequency of trivalent formation was considerably below the two-thirds expected on theoretical grounds.[36]

In later stages of pachytene in the triploid *Bombyx* occyte[32] the number of homologous trivalents present drops to 0, and 28 homologous bivalents are seen. The close alignment of the third member of each homologous trio is lost as the bouquet disappears and chromosomes become detached from the nuclear envelope. This stage is characterized by apparent nonhomologous SC formation between the unpaired axial elements to form heterologous bivalents, trivalents, and foldback hairpins. Chromosome pairing at pachytene in this triploid never reaches more than 80% completion.[32]

Rasmussen[32] interpreted this behavior in the triploid female as an example of two-phase pairing. The first phase is strictly homologous, while the second phase allows nonhomologous SC formation between unpaired regions or lateral elements. Von Wettstein et al.[37] call this second phase a "correction phase", which optimizes pairing. The absence of crossing over in female *Bombyx* allows the correction of trivalent pairing into homologous bivalents and the unpaired and displaced axial elements pair nonhomologously at this stage. The diploid achiasmatic female *Bombyx* ensures regular anaphase I disjunction and fertility by retaining modified SCs to hold bivalents together until metaphase I. This mechanism, although presumably present in the triploid females, cannot effect regular disjunction since some SCs hold together nonhomologous chromosomes. Hence triploid female *Bombyx*, like the chiasmatic triploid males, are sterile.

A somewhat different picture of pachytene pairing emerged from a serial section study of the triploid fungus *Coprinus cinereus*.[33] Throughout pachytene the pairing was almost completely as homologous trivalents, with no apparent alteration in pairing arrangement, even at early diplotene. Much of the pairing involved three homologous lateral elements and two central regions combining to form a "double SC" (triple pairing). The three lateral elements either lie side-by-side in one plane or form a three-sided prism. In the latter case, only two central elements were ever seen, never three. In nine reconstructed nuclei, about half of the trivalents had double SCs from end to end and a further 30% had one chromosome arm completely paired as a double SC and the other with one normal SC and one unpaired axial element.[33] Of the remaining trivalents, about half were triple-paired for part of one arm. Only about 10% of triploid *Coprinus* trivalents presented the conventional picture of two paired lateral elements and one unpaired axial element with a switch of pairing partner in the middle of the trivalent. Unpaired axial elements often had central region material attached and, in almost all trivalents, all three centromeres were triple paired.[33]

By using an electron-microscopic spreading technique, Loidl and Jones[34] have been able to analyze the SC formation in an autotriploid plant, *Allium sphaerocephalon*, with a haploid complement length at pachytene of approximately 600 μm. Individual chromosomes average about 70 μm in length,[34] compared with 2 to 4 μm in *Coprinus*,[33] and 3.5 to 6 μm in *Bombyx*.[32] It is apparent that trivalent pairing in the longer plant chromosomes, while having similarities with the previous two studies, also has some distinctive characteristics. Loidl and Jones[34] noted that, at late zygotene, two of the three axes (lateral elements) of each homologous chromosome trio were synapsed into SCs, and the third lay aligned with them, usually held at 200 to 600 nm by a number of association sites (Figure 1). They adopted the term "distance pairing" for this alignment of the third chromosome axis.[24,38] There were usually from 20 to 30 association sites spread unevenly along each trivalent, with few in the pericentromeric regions.[34] In the majority of trivalents, the telomeres of the three axes were connected by a spherule which persists well into pachytene. At mid-pachytene, when

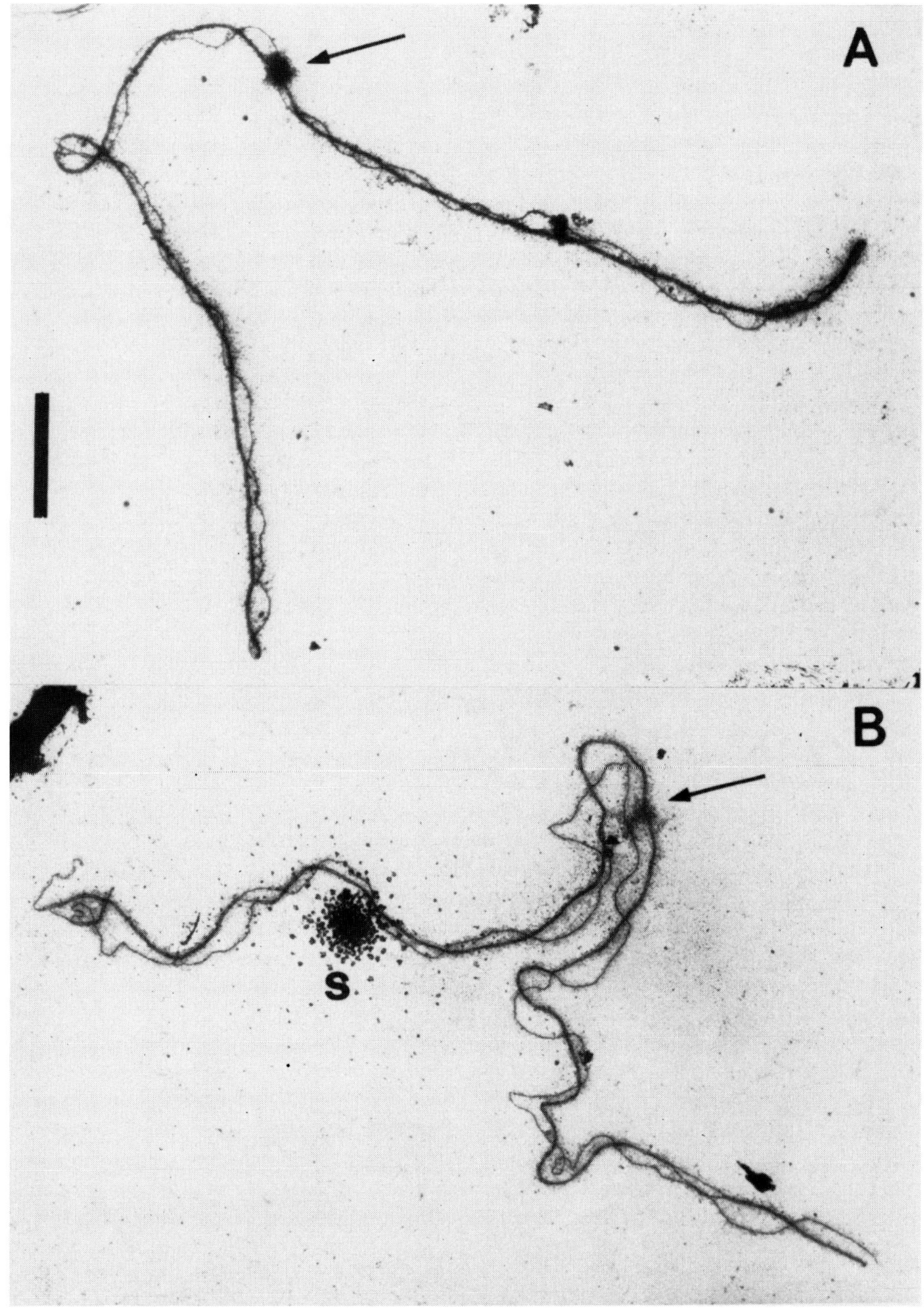

FIGURE 1.   Electron micrographs of spread trivalents from early pachytene (A) and mid-pachytene (B) micro-sporocytes of *Allium sphaerocephalon,* illustrating the decrease in number of intercalary associations of the three lateral elements during pachytene. Arrows indicate centromeres, S is a staining artifact. (Bar = 10 μm.) (From Loidl, J. and Jones, G. H., *Chromosoma,* 93, 240, 1986. With permission.)

distance pairing becomes less pronounced, Loidl and Jones[34] found conventional trivalents with one SC and an unpaired axial element at any site, with switches of lateral element pairing partners along the trivalent. An average of 3.4 switches of pairing partner occurred per trivalent (range 1 to 8; most had 2 to 4), so that only some of the zygotene associations must have involved pairing partner switches. No cases of triple pairing were reported in more than 30 *A. sphaerocephalon* nuclei examined.[34] As pachytene progressed, the unpaired lateral elements engaged in several forms of nonhomologous synaptonemal complex formation, both intrachromosomal foldbacks and interchromosomal associations. Pairing was maximized by the end of pachytene to form a network of interconnected synaptonemal complexes. Crossing over must be confined to homologously paired regions or occur prior to this nonhomologous pairing, since Loidl and Jones[34] found that >60% of the diplotene/metaphase I cells had seven or eight trivalents, indicating a high degree of homologous crossing over.

Loidl and Jones[34] interpreted their findings as indicating a high degree of alignment of homologous chromosomes during or prior to zygotene, with ''distance pairing'' associations providing a framework for formation of SC's between homologs. The latter could provide potential pairing sites, only some of which initiate SC formation. The number of pairing partner switches in a triploid should be related to the number of pairing initiation sites. Loidl and Jones[34] estimate a mean of 6.1 pairing initiation sites per trivalent in triploid *A. sphaerocephalon*. This is obviously much higher than the estimates of one pairing initiation at each telomere in *Bombyx*,[32] and illustrates how the length of chromosomes could be important in determining the pairing behavior of SCs in multivalents.[39] Alignment of axial elements by distance pairing, together with multiple initiations of SC formation, may be important in long chromosomes since zip-up pairing from telomeres would be both time consuming and more likely to lead to interlocks. The number of pairing partner switches is probably also of importance in determining the subsequent metaphase I multivalent frequency since the frequency of crossing over (chiasmata) does not rise in direct proportion to increases in chromosome size, but partner switches may reduce chiasma interference.[40] The large amount of DNA associated with plant chromosome axial elements probably can account for the elimination of distance pairing as pachytene progresses — contraction of chromatin around lateral elements will presumably increase the physical difficulties of these associations being maintained unless SC pairing at a site had produced a pairing switch to hold all three axial elements together.

These three autotriploids (*Coprinus*,[33] female *Bombyx*,[32] *A. sphaerocephalon*[34]) provide examples of three radically different behaviors in pachytene pairing. In triploid *Coprinus*, the short, low DNA content chromosomes are capable of complete triple pairing, probably from single initiation sites near one telomere of each chromosome, and facilitiated by the small amounts of chromatin associated with each axial element.[33] In female *Bombyx*, triploid pairing begins at both telomeres, but generally results in complete synaptonemal complexes involving only two of the three axial elements.[32] Because of the absence of crossing over, a second phase of nonhomologous pairing corrects most trivalent pairing into synaptonemal complexes involving only two lateral elements, albeit nonhomologous in some cases. In *A. sphaerocephalon* triploids,[34] the long chromosomes of each trisome associate (prealign) closely at zygotene and multiple random synaptonemal complex initiations result in a number of pairing partner switches per trivalent. However, all SCs involve only two lateral elements, presumably because of physical constraints imposed by the amounts of chromatin surrounding each. A second phase of nonhomologous pairing in this case involves only unpaired axial element segments from trivalents and, because of either distances involved or prior crossing over, does not result in displacement (resolution) of trivalent pairing.[34] In all three cases, the presence at metaphase I of a high frequency of trivalents (*Coprinus, A. sphaerocephalon*) or of nonhomologous paired bivalents and univalents (female *Bombyx*) is expected to result in unbalanced anaphase I segregation and sterility.

B. Triple Pairing

Using an early water-spreading technique, Comings and Okada[35] demonstrated that some of the pachytene trivalents in triploid chicken ovotestes, consisted of double SCs with three lateral elements associated side-by-side and joined by two central regions along all or most of their lengths. Up to five such triple paired trivalents were seen in one nucleus. This triple pairing appeared to be an exception to the accepted rule that only two axial elements could pair to form a SC at any one point. However, more recent studies have revealed that such triple pairing, while generally the exception, is by no means a rarity.

The most extensive illustration of its occurrence was in the serial-section analysis of triploid *C. cinereus* by Rasmussen et al.,[33] where it was found that the majority of pachytene chromosomes were present in trivalents with some triple pairing (see above). In this case, as in triploid chicken, trivalents could have double SCs from telomere to telomere. Other examples of triple pairing have been described in polyploids and trisomic chromosomes. Rasmussen[32] illustrated two examples of incomplete triple pairing near the telomeres of trivalents at zygotene in a triploid silkworm and a similar structure was recognized in a serial reconstruction of a trivalent from a female tetraploid silkworm.[41] In a surface-spreading study of SC formation in male tetraploid silkworms, Rasmussen[42] has found that triple and even quadruple pairing is not uncommon in stages of pairing prior to full pachytene. In the sectioning studies of polyploid silkworm,[32,41] the central region was not fully elaborated in one of the two SCs formed by triple pairing.

Electron-microscopic surface-spreading techniques have also revealed the occurrence of multiple paired SCs in mammals and plants. Wallace and Hultén[43] found that at pachytene in oocytes from a trisomy 21 (Down syndrome) fetus, 40% of nuclei had a trivalent present, the remainder having 23 bivalents and one univalent; 5 out of 36 trivalents had a double SC along their entire length; others had portions triple paired. Speed[44,45] has observed similar pairing of human chromosomes 18 and 21 in human aneuploid fetal prophase I (see also this volume. Chapter 1, Figure 6 and Chapter 2, Figure 8). In an electron micrograph of a spread of a spontaneously occurring trisomic spermatocyte from a bull, Dollin and Murray[46] showed a trivalent SC with triple pairing (Figure 2). Stack,[47] in a paper outlining a new spreading technique for plants, illustrates what appears to be an example of quadruple pairing in SC at the telomeres of a tetraploid potato quadrivalent. Gillies et al.[48] showed a short stretch of double SC at the telomeres of three homologs in an incompletely paired quadrivalent of autotetraploid *Triticum monococcum*. The somewhat unusual SC pairing described in tetraploid yeast[49] and tetraploid *Allomyces*[50] are possibly further examples of triple or quadruple pairing.

The demonstration of triple (and quadruple) pairing in SCs of autopolyploids and trisomics confirms what has been suggested by previous light-microscopic studies of pachytene stages. A number of workers have observed that light microscopy was unable to distinguish the three chomosomes closely associated in parts of trivalents.[51] In part, this may be due to an inability of the light microscope to distinguish between true triple pairing and distance pairing of an unsynapsed axial element beside a normal SC.[24,34]

Pairing of one lateral element with two others raises some questions about the concept of SC formation. Early light-microscopic studies of meiotic chromosome pairing established that only two homologs usually associated at any one point.[52-54] Electron microscopy of SC formation, in general, confirmed this situation[21,22] and the view has prevailed that, during zygotene synapsis, only two axial elements can associate to form an SC. Von Wettstein[55,56] has proposed a model in which the axial element at lepotene/zygotene becomes positioned lateral to the chromatin from the two sister chromatids of a chromosome. Side-by-side pairing can then occur between axial elements from two homologs to form one SC. Support for a single-pairing face in lateral elements came from the finding by Moses et al.[57] that the kinetochores of the two acrocentric chromosomes in *Lemur* Robertsonian translocation hybrid trivalents were always on the same side of the SC (i.e., in a *cis* configuration).

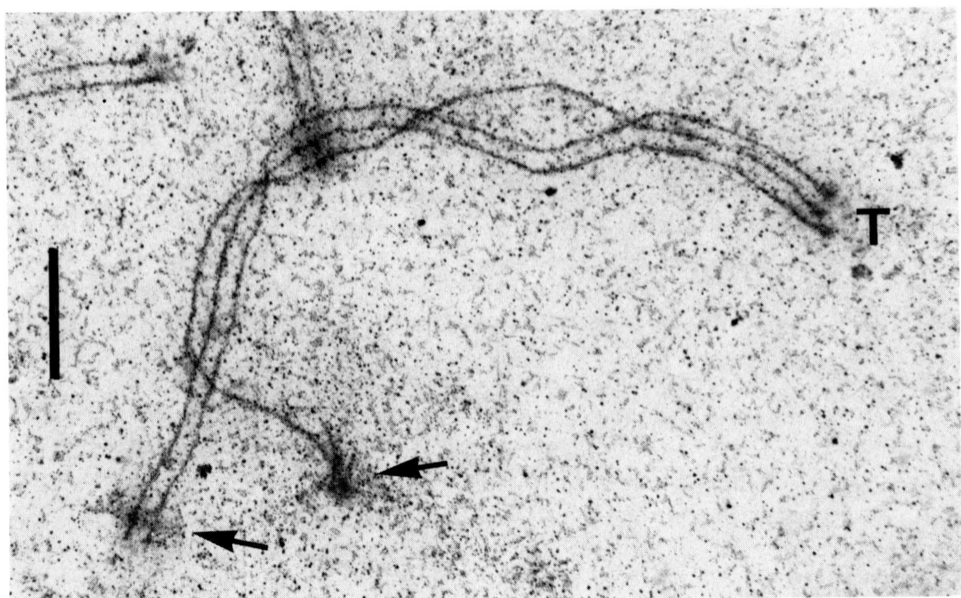

FIGURE 2.   Electron micrograph of a spread trivalent with double synaptonemal complex in an aneuploid spermatocyte from a Brahman X Hereford F1 bull. T = triple paired telomere, arrows indicate centromeres. (Bar = 1 μm.) (Unpublished original from A. Dollin and J. Murray. With permission.)

However, reports of "polycomplexes", stacks of multiple SCs, have demonstrated that in certain situations this constraint on pairing does not hold.[58] There are a number of possible factors which may normally restrict pairing to formation of a single SC, and some of these may be absent in the cases of triple pairing. The majority of cases of triple pairing involve either organisms with short chromosomes (*Coprinus*, yeast, *Allomyces*) or only short chromosomes or chromosome segments (chicken, humans, cattle, *Bombyx*, *T. monococcum*, potato). This suggests that there may be physical constraints on triple pairing over long distances. Conversely, partner switches necessary for normal SCs in trivalents may require greater lengths than are available in short chromosomes. Rasmussen et al.[33] suggested that low amounts of DNA associated with each lateral element in the triploid fungus *Coprinus* might account for the ease with which they form triple-paired trivalents. In higher organisms, the extensive chromatin associated with the middle lateral element of a double SC would need to be displaced during triple pairing, and this might not be possible over any great distance. The examples of short telomeric stretches of multiple pairing (*Bombyx*, *T. monococcum*, potato) also suggest that close alignment of all three lateral elements (as might be expected if pairing is initiated at or near telomeres) is a prerequiste which may rarely be achieved. In tetraploids, a combination of the drive to maximize pairing and the arrangement of chromatin and lateral elements probably almost always leads the four homologous lateral elements into forming two SCs. Only when one lateral element is misaligned or trapped elsewhere would three homologous lateral elements be available and alter their chromatic configuration to allow commencement of triple pairing. In a tetraploid *Bombyx* spermatocyte nucleus from an animal showing arrested pairing, Rasmussen[42] found nine examples of triple and quadruple pairing at telomeres.

The fungi, with their low DNA contents, show the most extensive associations of more than two lateral elements, and may be considered examples of a more primitive type of pairing. Bojko[59] has recently described examples of triple pairing in diploid *Neurospora*, in some cases associated with abnormal split lateral elements. The reason for this triple pairing was unclear. There was no apparent effect on fertility of ascospores. With the increase in

DNA content in higher organisms, the associations between three lateral elements in a trivalent are physically constrained by the chromatin, and the phenomenon of distance pairing may be considered the successor to triple pairing in chromosomes with high DNA contents.[24,34] In abnormal pairing situations such as animal triploids and trisomics,[35,43,46] short chromosomes with low DNA contents may still be capable of triple pairing over a few microns from telomeres. The significance of triple pairing for crossing over and fertility may be minor if it is resolved during pachytene.[42] Whether both SCs in a triple-paired region are competent for crossing over[59] is uncertain. Recombination nodules were observed in both SCs of triple-paired *Coprinus* trivalents, although they occurred independently and their total frequency in the triploids was the same as in the diploids.[33]

C. Later Stages — Light-Microscopic Data

Considerable data exist on the metaphase I/anaphase I behavior of autotriploid plants, and on their fertility and breeding behavior. Kuspira et al.,[60] in a study of autotriploid *T. monococcum*, reviewed a large number of reports. They found that the frequencies of trivalent formation at metaphase I in five x = 7 plant species ranged from 0.38 to 0.88. In two x = 12 autotriploids, the trivalent frequencies were 0.81 (*Lilium tigrinum*) and 0.41 (*Lycopersicon esculentum*). They suggested the difference between these two could be due to the longer chromosomes in the lily, but within the x = 7 species there was no obvious relationship between trivalent frequency and chromosome length. They concluded that genetic factors must also be involved in the determination of metaphase I behavior. Obviously, factors such as chiasma frequency and position could influence metaphase I pairing behavior, in addition to the interaction of chromosome length, pairing partner switches, and chiasma interference mentioned previously. Some species with long chromosomes can have a high trivalent frequency at pachytene and sufficient chiasmata dispersed on chromosome arms to maintain this pairing until metaphase I.[34] The regular occurrence of crossing over in most plants makes it unlikely that correction of trivalent pairing (as seen in female *Bombyx*) would be important in reducing trivalent frequencies from pachytene to metaphase I. In autotriploids, the occurrence of a bivalent + a univalent at metaphase I is the result of failure of either pairing (asynapsis) or crossing over (desynapsis) to involve one chromosome of a homologous trio. An excess of univalents is presumably the consequence of asynapsis or desynapsis of all three homologs, usually a rare event.[60] Distinguishing between pairing- and crossing-over effects can be difficult from metaphase I data and pachytene pairing may be difficult to analyze in many triploids. This question will be discussed in more detail in conjunction with autotetraploids.

If trivalent frequency is high, the metaphase I orientation of the trivalents will be a determining factor in anaphase I segregation. Kuspira et al.[60] have recently shown that in autotriploid *T. monococcum* metaphase I cells with 3, 4, or 5 trivalents, the orientation of trivalents was random. Since univalents usually segregate randomly at anaphase I, and bivalents segregate regularly in a 1:1 manner, then the majority of telophase I nuclei should have approximately equal number of chromosomes, i.e., a number midway between x and 2x. In species where aneuploidy is not well tolerated, this results in a drastic reduction in fertility in autotriploids. In triploids such as bananas, where irregular segregation or complete breakdown of anaphase I leads to the absence of any pollen,[61] vegetative propagation allows the triploid to persist. Kuspira et al.[60] found that 80% of telophase I nuclei in triploid *T. monococcum* (3x = 21) had 9 to 12 chromosomes. Fewer than 1% of the cells had x or 2x chromosomes. Selfing triploid *T. monococcum* resulted in a seed set of only 0.23%[60] and 94% of the resultant progeny were 2x or 2x + 1. In a survey of 13 species in which 3x females had been crossed to 2x males, Kuspira et al.[60] found that, in 8 of the species, more than two thirds of the progeny were 2x or 2x + 1. Four species (*Clarkia unguiculata, Oryza sativa, Petunia hybrids, Zea mays*) seemed more tolerant of aneuploidy and had significant proportions of 2x + 2, 2x + 3, and 2x + 4 progeny.[60] Triploid maize gave a seed set of 11% in 71 ears from 56 plants.[62] Thus, it appears that tolerance of additional chromosomes can be an important factor in determining fertility of triploids, and will act independently of any meiotic effects.[60]

III. AUTOTETRAPLOIDS

A. Prophase I Light-Microscopic Data

More than 50 years ago, Darlington[53,63] described the pachytene pairing in tetraploid *Hyacinthus* and *Primula*, and the exchange of pairing partners in the four homologues to form quadrivalents. Levan[64] subsequently found that most quadrivalents at pachytene in autotetraploid *A. porrum* (a species with proximally localized chiasmata) had only one pairing partner switch. There was always at least one quadrivalent per nucleus at pachytene (and sometimes up to eight) and they were still readily seen at diplotene. However, during diakinesis and metaphase I, the majority of chromosomes were present as bivalents having two proximal chiasmata, although a few bivalents had only one chiasma each. Quadrivalents were present in only a small proportion of metaphase I cells and, again, the sites of chiasmata were proximal to the centromeres. Levan[64] concluded that if the majority of pachytene quadrivalents had a single switch of pairing partner which occurred outside the pericentromeric region where crossing over occurs, the result would be the formation at metaphase I of two bivalents, each with two proximal chiasmata. If the pairing partner switch occurs in the pericentromeric region, but to one side of the centromere, it may interfere with crossover events at that site and the metaphase I result would be either two bivalents with one chiasma each or a chain quadrivalent. Only if the pairing partner switch occurs at or very close to the centromeres would a ring quadrivalent result at metaphase I. Only three such cases were seen in many hundreds of metaphase I cells scored.[64]

Since these early studies, only a few other light microscope studies have examined the occurrence and consequences of pachytene pairing in autotetraploids. Gottschalk[65] described associations of the four NOR chromosomes of tetraploid *Solanum lycopersicon* at zygotene and pachytene, some of which involved quadrivalents with single pairing-partner switches. In other cases, he observed loose pairing between whole or parts of two NOR bivalents, which he termed "secondary pairing (Sekundäpaarung)". Gottschalk[66] notes that there is no substantial connection between the secondarily paired homologous regions of the two bivalents, but the association may be the residual effect of primary pairing forces. This behavior has similarities with the distance pairing described in triploids.[34]

Subsequent light-microscopic studies of pachytene in autotetraploid plants have often concentrated on karyotypic details,[67] but some studies have attempted to analyze the nature of pachytene pairing and partner switching. In a tetraploid race of *Sorghum arundinaceum*, the pachytene quadrivalent frequency was inversely proportional to chromosome length,[68] with the smallest chromosomes of the complement usually present at pachytene as bivalents. The longer chromosomes paired as quadrivalents up to 60% of the time, mostly having single pairing-partner switches, although the largest chromosome often had quadrivalents with 2 or 3 switches. Only the longest chromosomes approached the theoretical expectation[36] of two thirds quadrivalent pairing and, overall, 45% of the pachytene chromosomes were present as quadrivalents, but only 36% were found as quadrivalents at diakinesis.[68] In the presumed autotetraploid *Medicago sativa* (2n = 4x = 32), there were 0.89 to 2.93 quadrivalents at pachytene, but dissociation of quadrivalents because of low chiasma frequency reduced this to 0.36 to 1.32 quadrivalents per metaphase I cell, together with a few trivalents and univalents.[69] The initial low pachytene quadrivalent frequency (well below the theoretical two thirds) suggests that a control on pachytene pairing is involved.

Very few light-microscopic studies of animal polyploids at early prophase I have been reported. John and Henderson[40] used the occurrence of occasional spontaneous tetraploid cells in the meiotic stages of the male locust *Schistocerca paranensis* to analyze aspects of multivalent formation at diplotene and later. They concluded that "quadrivalent formation is predominantly a function of relative chromosome length and of chiasma frequency".[40] At diplotene/diakinesis, the long chromosomes formed multivalents 93% of the time, the medium chromosomes 74% of the time, and the small chromosomes only 17% of the time. They also found the chiasma frequency higher in tetraploid cells than in diploids, and attributed this to partner switches preventing chiasma interference. The high frequency of quadrivalent formation in long chromosomes presumably was the consequence of several points of pairing initiation, and a mean chiasma frequency of greater than two per chromosome pair. John and Henderson[40] also explained the low frequency of trivalents and univalents with a model of random, but obligatory, pairing of the four homologs so that no chromosome remained unpaired by the end of pachytene. Beçak et al.[70] briefly studied male meiosis in tetraploid and octaploid frogs. In the tetraploid, they found terminal pairing in a bouquet at zygotene, with incomplete interstial pairing in quadrivalents. Some bivalent pairing was apparent at zygotene, and trivalents and univalents were rare at metaphase I. Beçak et al.[70] also noted octavalents, hexavalents, quadrivalents, and bivalents at metaphase I in the octaploid species. Detailed metaphase I analyses were subsequently done by Rahn and Martinez.[8]

The majority of light microscopic studies of pachytene pairing in autotetraploid species (both plant and animal) have been hampered by technical problems which precluded analysis of the completeness of pairing, both in individual homologous sets, and in whole nuclei. The recognition that at zygotene/pachytene, chromosome pairing can be monitored by electron microscopic studies of axial elements/synaptonemal complexes has allowed further studies of the nature of autotetraploid chromosome pairing.

B. Prophase I Electron-Microscopic Data

In the first serial section analysis of SC formation in an autotetraploid, Moens[28] traced four different quadrivalent pairing-partner switches in the autotetraploid *Lilium longiflorum* pachytene microsporocytes. In all cases, the four homologous axial elements were paired in two SCs at one site and switched to a different combination of axial element (AE) pairs in two synaptonemal complexes at another site, i.e., AEs 1 + 2 and 3 + 4 switch to 1 + 4 and 2 + 3. Thus, the axial elements (lateral elements) represent the four homologous chromosomes and behave as expected according to previous light microscope studies of pachytene and later meiotic stages.

Studies of serially sectioned meoitic nuclei from tetraploid strains of the fungi *Saccharomyces cerevisiae*[49] and *Allomyces macrogynus*[50] have revealed somewhat equivocal evidence about quadrivalent SC formation. In tetraploid yeast, Byers and Goetsch[49] found close side-by-side associations of pairs of SCs with regions of indistinct synaptonemal complex which were consistent with pairing-partner switches between lateral elements, as expected if quadrivalents were formed. On the basis of observation of apparently double central components in some synaptonemal complexes, Borkhardt and Olson[50] postulated that quadrivalent pairing occurred in incompletely paired chromosomes of the tetraploid *A. macrogynus*. Later in pachytene, only bivalents were observed, and they concluded that dissolution of the abnormal SCs allowed normal central regions to assemble exclusively bivalent SCs.

SC formation at pachytene predominantly or even exclusively as bivalents, has been reported in serial section studies of the autotetraploid female *Bombyx*[41] and in several autotetraploid nematodes.[12,13] At early pachytene in the female tetraploid *Bombyx*, Rasmussen and Holm[41] found from 5 to 12 quadrivalents (mean 8.4), 27 to 44 bivalents (mean 36.7), with only occasional univalents and trivalents (Figure 3). The majority of quadrivalents had single pairing switches, while a few had two. Examples were seen of nonhomologous

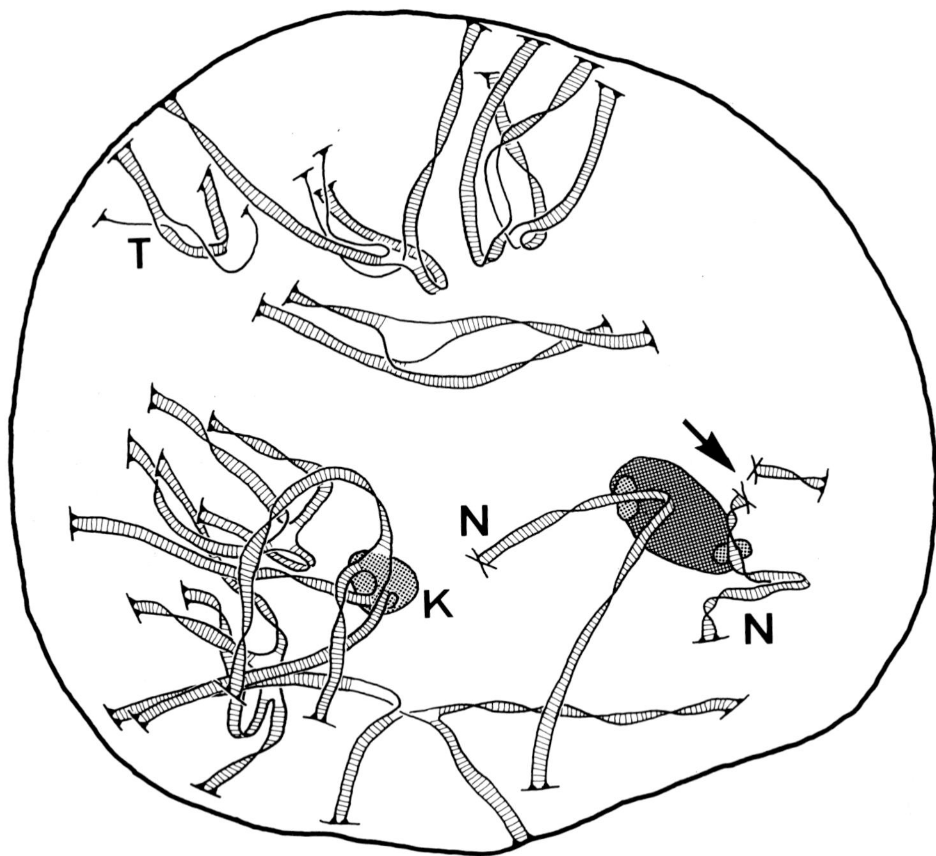

FIGURE 3.   Partial reconstruction of serial electron micrographs of an early pachytene nucleus from a female *Bombyx* autotetraploid showing seven quadrivalents and one trivalent (T), together with the two NOR bivalents (N) associated with the nucleolus (dark hatching). All telomeres are attached to the nuclear envelope, except for one NOR bivalent (denoted by cross), and one NOR bivalent is broken (arrow). A chromatic knob (K) is associated with one quadrivalent. (From Rasmussen, S. W. and Holm, P. B., *Carlsberg Res. Commun.*, 44, 101, 1979. With permission.)

associations in parts of multivalents. However, in 11 late pachytene stage nuclei there were only 0 to 2 quadrivalents, and from 48 to the maximum of 56 possible bivalent synaptonemal complexes (mean 51.9). Once again, trivalents and univalents were rare. Unlike zygotene/ early pachytene, the ends of SCs were not attached to the nuclear envelope, and no bouquet was recognizable. The reduction in quadrivalent number between the early and late pachytene was interpreted by Rasmussen and Holm[71] as a correction process to maximize bivalent pairing, as also seen in the triploid female *Bombyx*.[32] The results from the triploid demonstrate that correction pairing does not appear to depend on strict homology. In the case of the tetraploid, correction results in only bivalents at metaphase I and hence, regular anaphase I segregation and fully fertile gametes. Such reorganization of quadrivalent pairing into bivalent pairing at late pachytene is possible because of the absence of crossing over in the female *Bombyx*. In the male tetraploid *Bombyx*, crossing over precludes correction[42] (see following discussion).

The pachytene pairing in tetraploid nematodes is more difficult to interpret. In four early to mid-pachytene nuclei reconstructed from autotetraploid *Meloidogyne hapla* females, Goldstein and Triantaphyllou[13] found exclusively 34 bivalent SCs in each nucleus, which was in agreement with the 34 bivalents seen at metaphase I. In another tetraploid nematode,

Heterodera glycines, these authors also found only bivalent SCs in three pachytene nuclei,[13] even though metaphase I cells with quadrivalents, trivalents, and univalents had been reported, in addition to others with all bivalents.[72] Two interpretations of the *H. glycines* results are possible. Either the sample of three pachytene cells examined by electron microscopy simply failed, by chance, to include nuclei with multivalent pairing, or the interpretation of the metaphase I nuclei[72] was at fault in reporting associations as trivalents and quadrivalents. The results from *M. hapla* suggest the latter may be the correct view, in which case the two autotetraploid nematodes show exclusive bivalent pairing, which may be the result of each chromosome having a single nuclear envelope attachment and pairing initiation site.[13] Pairing which began at a single site between two homologs would most likely proceed to complete bivalent synaptonemal complex formation without the opportunity for partner switches.

The advent of EM surface-spreading techniques has made easier the analysis of synaptonemal complex behavior in nuclei from autotetraploids. Solari and Moses[73] demonstrated that, in a spontaneously occurring tetraploid pachytene spermatocyte from mouse, it was possible to recognize six definite and three possible quadrivalents, as well as 20 bivalents. No trivalents were seen. Most quadrivalents had a single-partner switch. The two X and Y chromosomes each formed unpaired, thickened axial elements.[73] Stack,[47] in a paper describing a new synaptonemal complex spreading technique for plants, demonstrated that autotetraploid potato at pachytene could have 9 or 10 quadrivalents, usually with single-partner switches. Triple and quadruple pairing was possible near telomeres, and nonhomologous synaptonemal complex formation indicated either structural rearrangements or mismatch and foldback pairing. Almost all ends of chromosomes were fully paired in both bivalents and quadrivalents.[47] Spreading analysis of synaptonemal complexes is particularly relevant to plants, where the mechanics of three-dimensional reconstruction of serial sections makes analysis of more than a few zygotene/pachytene nuclei virtually impractical. The full worth of EM spreading has begun to be realized with recent papers reporting on synaptonemal complex behavior in both auto- and allopolyploids.

Following a study of the triploid *Allium sphaerocephalon*,[34] Loidl[74,75] has spread zygotene/pachytene nuclei from a number of *Allium* species, including the autotetraploid *A. vineale* (2n = 4x = 32), and the autopentaploid *A. oleraceum* (2n = 5x = 40). At early zygotene in the tetraploid *A. vineale*, he found axial elements aligned in bundles of four (presumably homologous tetrasomes) along much of their lengths.[74] The alignment appeared to be facilitated by up to 10 intercalary pairwise associations of axes per tetrasome. SC formation initiated in subterminal regions, not at telomeres. By late zygotene, two synaptonemal complexes were formed along most of the length of each tetrasome, with frequent switches of pairing partner.[74] Terminal associations also connected homologous telomeres via nuclear envelope attachments. During pachytene, the alignment of axes within tetrasomes disappears, except for partner switches, and each tetrasome can be seen as a quadrivalent SC, or, occasionally, as a pair of bivalent synaptonemal complexes. Of 85 tetrasomes examined, 17 were present as two bivalents. Of the remaining 68 quadrivalents, 33 had one partner switch, 21 had two switches, 10 had three switches, 3 had four switches, and 1 had five switches.[74] There was a mean of 1.4 switches per tetrasome. Using the number of switches, Loidl[74] estimated the number of initiation sites per tetrasome as 6.2 (i.e., 3.1 independent initiations since they occur in nonindependent pairs). The proportion of quadrivalents at pachytene was 80% (68/85) and at diplotene-metaphase I, 22% (52/240). Many of the metaphase I bivalents were rod shaped, indicating a low chiasma frequency, which could explain some of the reduction in quadrivalent frequency from the higher pachytene value. Loidl[74] also speculated that chiasma localization or crossing over preferences among the tetrasomes might also favor bivalent formation at metaphase I.

In a brief study of pentaploid *A. oleraceum*, Loidl[75] found axial elements aligned at

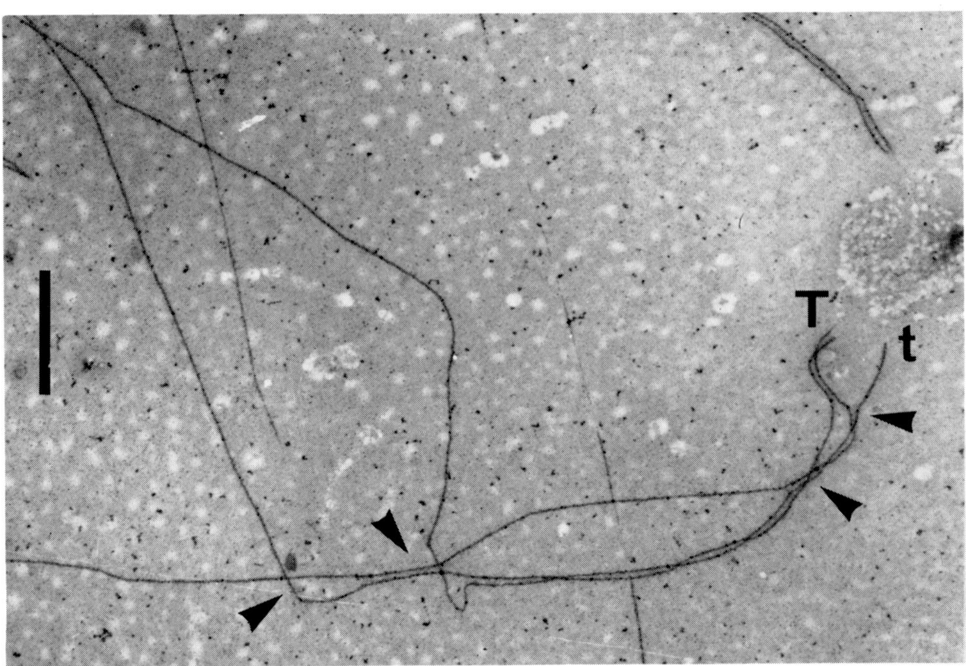

FIGURE 4.   Electron micrograph of a spread trivalent from an autotetraploid *Triticum monococcum* microsporocyte showing several synaptonemal complex pairing partner switches (arrowheads). Paired (T) and unpaired (t) telomeres are indicated. (Bar = 2 μm.) (From Gillies, C. B., Kuspira, J., and Bhambhani, R. N., *Genome*, 29, 309, 1987. With permission.)

leptotene in bundles of five, which subsequently become paired at zygotene into two synaptonemal complexes with one axis left unpaired. Terminal associations of axes were again common. As in the tetraploid and triploid *Allium* species,[34,74] intercalary associations of axes and switches of pairing partners were seen. However, pairing was never seen to be complete. This may explain the low multivalent frequency at diakinesis-metaphase I (0 to 1 pentavalents, 1 to 2 quadrivalents, 2 to 3 trivalents, 4 univalents, and the rest bivalents[76]). Pollen fertility was only 5 to 10% and male gametes were unbalanced, with chromosome numbers in microspores ranging from 17 to 24. In spite of this, Tschermak-Woess[76] found good seed set and seed germination, suggesting that apomixis operated.

Gillies et al.[48] used a spreading technique to analyze SC formation in autotetraploid *T. monococcum* (2n = 4x = 28). In 22 mid-zygotene to pachytene stage nuclei, they found a minimum of 1 to 6 multivalents per nucleus (mean value 3.59 per nucleus). Almost all multivalents were quadrivalents, with 75 out of 89 having a single pairing-partner switch, 7 having two switches, 4 having three switches, and 3 having four or more switches (Figure 4). At mid-zygotene, switches were often associated with lateral element asynapsis. The mean number of switches was 1.27 per multivalent,[48] compared with a mean of 1.79 switches per quadrivalent SC in *A. vineale*.[75]

The zygotene/pachytene multivalent frequency in *T. monococcum* (minimum mean 3.59 per nucleus) was less than the theoretical two thirds frequency[36] and the distribution of a number of multivalents per cell did not fit a binomial expectation.[48] The multivalent frequency in this study[48] was at least double that found at metaphase I in the same material,[77] but much closer to the mean value of 5.1 quadrivalents per metaphase I cell found in another line of autotetraploid *T. monococcum* by Morrison and Rajhathy.[78] It was suggested by Kuspira et al.[60] that genotypic differences between the two *T. monococcum* lines might

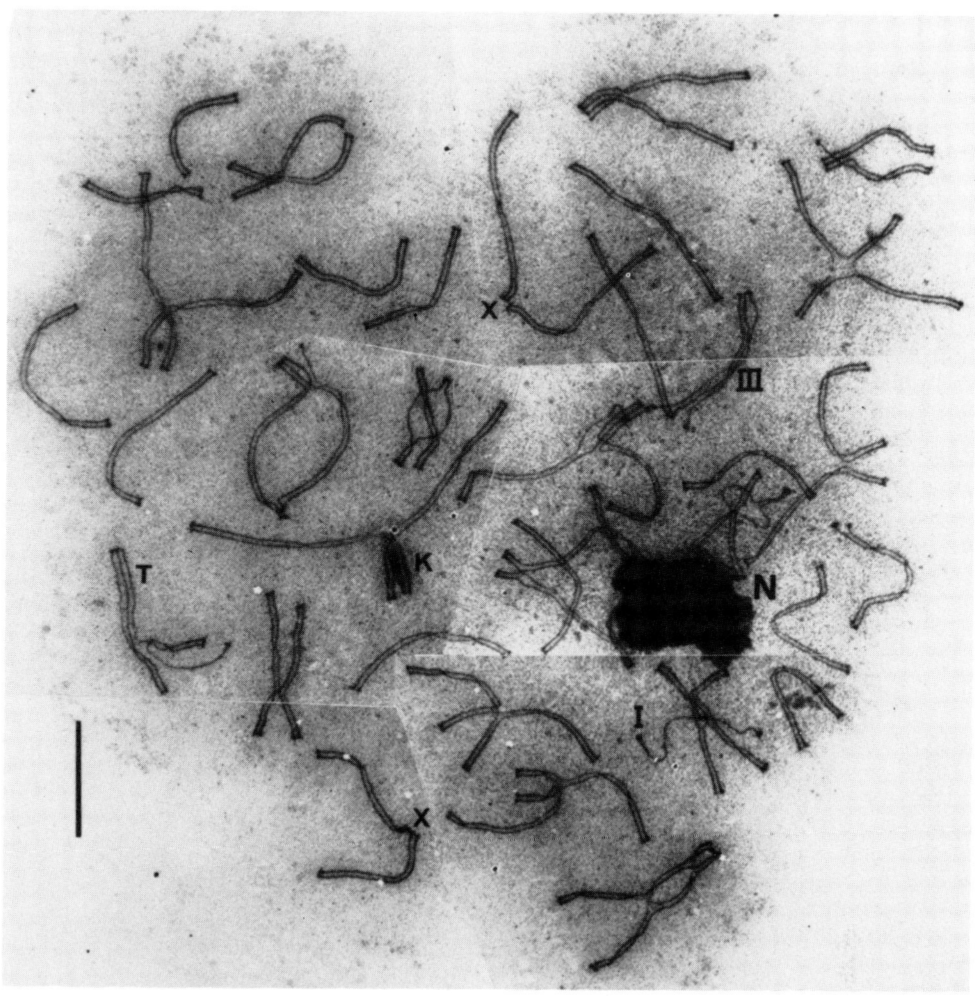

FIGURE 5.   Electron micrograph of a spread early pachytene (95% paired) complement from an autotetraploid *Bombyx* with 16 quadrivalent synaptonemal complexes, one trivalent (III) and univalent (I), and 22 bivalents. The knob associated with chromosome 1 is visible (K). One quadrivalent shows triple pairing (T) and two pairs of presumed homologous bivalents are associated at their telomeres (X). N = nucleolus. (Bar = 5 μm.) (From Rasmussen, S. W., *Carlsberg Res. Commun.*, 52, 211, 1987. With permission.)

explain the difference in metaphase I quadrivalent frequencies. Factors which Gillies et al.[48] considered might be instrumental in reducing multivalent frequency from pachytene to metaphase I, included preferential pairing or crossing over, localized partner switches, and asynapsis.

Rasmussen[42] has recently published the most extensive analysis of pachytene pairing in an autotetraploid, an EM study of 142 spread nuclei from four autotetraploid male *Bombyx mori* (Figure 5). Measurements of lateral element length, percentage pairing, and number of univalents, bivalents, trivalents, and quadrivalents were made in approximately 30 nuclei at each of late zygotene, early pachytene, and mid to late pachytene. During zygotene, pairing followed the diploid pattern,[79] i.e., subterminal lateral element recognition and then SC formation in the region between the recognition site and the telomere, followed by subsequent interstial SC formation. In many cases at zygotene in the autotetraploid, three or four homologous lateral elements were aligned in near parallel arrangement over all or part of their lengths. In some cases, this was a consequence of triple paired regions of double

SC. In others, classical multivalent SC pairing was seen. In a few cases, two presumed homologous bivalents lay parallel or with attachment plaques associated (Figure 5).

Although most of the SCs in the tetraploid *Bombyx* could be classified by length alone, it appeared that all associations were homologous.[42] Interlocks were seen, but they averaged fewer than one per nucleus. They appeared to be one cause of univalency. Interlocks of homologs were also seen, which appeared to lead to two pairing-partner switches in quadrivalents. By the end of zygotene, all univalents were paired, and only bivalents and quadrivalents (no trivalents) were seen at pachytene.[42] No triple pairing persisted into pachytene. Pairs of presumed homologous bivalents were separated farther at pachytene than at zygotene. Analysis of pairing was complicated by the presence of translocations in some animals, and one of the four had pairing arrested at about 50% in all nuclei, together with circular lateral element fragments.[42] In this animal, irregular nonhomologous SCs were common, with up to 30 axial elements involved in large complex chains held together by short telomeric SC segments, including side-by-side multiple pairing of four lateral elements in some cases.[42]

In normal *Bombyx* autotetraploid nuclei, the frequency of multivalent SC formation decreased as pairing progressed.[42] At mid- to late-zygotene (see Figure 5), 7 to 18 quadrivalents were present per nucleus (mean of 13.3 in 33 nuclei), with 0 to 9 trivalents and univalents (means 0.7 and 1.0, respectively), and 15 to 39 bivalents (mean 25.1). By early pachytene, the mean values from 27 nuclei were 11.1 quadrivalents, 31.5 bivalents, 0.2 univalents, and no trivalents. In 27 mid- to late-pachytene nuclei, there were means per nucleus of 8.7 quadrivalents and 37.0 bivalents, with no trivalents or univalents. Random pairwise association of both telomeres is predicted to generate two thirds quadrivalents,[36] which in a 2n = 4x = 112 genotype (silkworm) would produce 18.67 quadrivalents per nucleus. The maximum number of quadrivalents in this case (observed at zygotene) was 18,[42] close to expectation, although the mean value was lower. The reduction in quadrivalent frequency during pachytene was interpreted by Rasmussen[42] as the result of progressive conversion of quadrivalents into pairs of bivalents. The pachytene pairing behavior in the *Bombyx* male tetraploids was almost identical to that seen at early pachytene in serially sectioned female tetraploids.[41] However, light-microscopic data (cited in Reference 80) showed that the tetraploid male has a mean of 6.7 quadrivalents and 42.2 bivalents at metaphase I, whereas the tetraploid female has almost all bivalents at metaphase I. Rasmussen[42] suggests that the final metaphase I quadrivalent frequency in the male tetraploid *Bombyx* is determined by an interaction between quadrivalent resolution into bivalent synaptonemal complexes and crossing over generating chiasmata which maintain quadrivalents (i.e., prevent their resolution). Rasmussen[42] serial sectioned two tetraploid male pachytene nuclei and found 50 and 58 recombination nodules, respectively, with a distribution within quadrivalents suggesting that crossing over at these sites would maintain quadrivalents until metaphase I. Thus, the critical factor in the male *Bombyx* is time of crossing over.

Rasmussen[42] suggests that regular bivalent presence at metaphase I in tetraploids can be the result of at least three types of behavior:

1. Zygotene pairing restricted to bivalent SCs due to simple pairing initiations per chromosome (e.g., tetraploid nematodes)[12,13]
2. Localized chiasma formation in pachytene quadrivalents (e.g., *A. porrum*)[64]
3. Complete resolution of zygotene quadrivalents into bivalents at pachytene (in the absence of, or prior to, crossing over, as in the female tetraploid *Bombyx*),[42]

C. Theoretical Models of Chromosome Pairing in Autopolyploids

Over the years, a considerable amount of time has been devoted to discussion of which are the important factors in determining chromosome associations at meiosis in polyploids. A number of models have been developed to predict the type of chromosome pairing expected at pachytene and/or metaphase I, taking into account such factors as numbers and sites of pairing initiation, and of crossover events. The predictions of these models have then been compared with real data from polyploids (usually light microscopic metaphase I data) to see whether the assumptions in the models accurately pairing events. Some of these models will be reviewed briefly and recent results from light and electron microscopy, primarily in teraploids, will be related to them.

John and Henderson[40] had noted the relationship of quadrivalent formation with chromosome length and chiasma frequency, and showed that in metacentric chromosomes with random terminal pairing initiations in both arms, the number of quadrivalents formed in a cell should equal the number of bivalents (i.e., two thirds of the chromosomes should pair as quadrivalents). Sved[36] reiterated this result and related the terminal pairing initiations to the telomeric attachment of chromosomes to the nuclear envelope. He noted that quadrivalent frequencies higher than two thirds might be explained by pairing initiation at more than two sites per chromosome, as found in maize.[36] John and Henderson[40] suggested that long chromosomes of tetraploid *Schistocerca paranensis* with more than two chiasmata, which formed quadrivalents 93% of the time, apparently possessed several pairing initiation sites.

Sybenga[81] developed this theory further by considering the effect on metaphase I pairing of the presence or absence of chiasmata in each of the two arms of metacentrics. He derived formulae allowing the use of metaphase I data to estimate the frequency of quadrivalent pairing at pachytene, as well as the comparison of chiasma formation in quadrivalents and bivalents. Applying these formulae to data from autotetraploid *Tradescantia virginiana*, he found that the estimated pachytene pairing was 66% quadrivalent formation, even though only 55% quadrivalents were present at metaphase I.[81] In addition, chiasma frequencies were different after quadrivalent and bivalent pairing.

Jackson and co-workers[82,83] have developed models for determining the frequencies of various metaphase I pairing classes in autotetraploids having 2, 3, or 4 chiasmata per bivalent, using only the basic chromosome number and the chiasma coefficient (average observed number of chiasmata divided by theoretically possible number). After correcting for the nonrandom distribution of chiasmata, they found that the predictions of the model fitted data from autotetraploids *Triticum longissimum* and *Haplopappus spinulosus*.[83] The uncorrected model[82] also gave predictions in agreement with John and Henderson's data[40] for autotetraploid *Schistocerca* chromosomes with 3 and 4 chiasmata per bivalent. It is suggested[82,84] that cases of deviations from expectations of the model could be the result of *Ph*-like gene effects, structural changes in chromosomes, or preferential pairing. Jackson[84] was of the opinion that gene mutations which alter the positions of genomes on the nuclear membrane were the major cause of alloploid-like behavior in autopolyploids.

A model relating pachytene multivalent frequency in both odd and even numbered polyploids to the number of "autonomous synaptic sites" — regions in homologs capable of producing a switch in pairing partner — was presented by Callow and Gladwell.[85,86] The model assumes pairwise synapsis occurring sequentially at a number of different sites on each chromosome set and with the same probability at each site. Pairing is assumed to be complete in even-numbered polyploids, but the equivalent of one chromosome set remains unpaired as univalents or segments in odd-numbered polyploids.[85,86] It was found that increasing the number of synaptic sites maximized the even-numbered multivalent frequency. However, multivalent frequency was also found to be a function of bivalent frequency, ploidy, and the minimum number of pairing-partner switches required to produce the multivalent.[85] Hamey et al.[39] have explored the implications of this model, including the effect of chromosome size, in autotetraploid barley and various *Crepis* polyploids. Data from 60 metaphase I cells of hexaploid *C. capillaris* fitted predictions of the model for both two discrete synaptic sites per chromosome and continuous synaptic sites.[39,86] Larger chromosomes formed more multivalents with a higher frequency of partner switches and chiasmata.

In autotriploid *C. rubra* and autotetraploid *C. capillaris*, the results showed that both the number and size of synaptic sites increased with the size of chromosomes.[39] There were more trivalents in 3x *C. rubra* than quadrivalents in 4x *C. rubra*, which was interpreted as evidence for more partner switches in the triploid, possibly because of slower pairing.[39,86] A similar effect was noted in a comparison of 3x and 4x pachytene pairing in *Allium*.[34,74] Evidence from electron microscopic analyses of zygotene pairing in maize,[87] lily,[88] *Tradescantia*,[89] and rye[90] suggests that in these diploid plants the number of pairing initiations per bivalent is usually many more than two.

D. Model Appraisal — Light-Microscopic Meiotic Data

There have been numerous light-microscopic studies of metaphase I pairing behavior in (supposed) autotetraploids (see Reference 77). In some cases, the theoretical two thirds frequency of quadrivalents was found (e.g., Morrison and Rajhathy[78]), but in many instances the quadrivalent frequency was less than that predicted by any of the models.[77,91-95] Factors which were suggested to explain these deviations from expectations included one or more of the following:

1. Genetically controlled preferential pairing
2. Structural differences in chromosomes leading to preferential bivalent pairing or crossing over
3. Reduced chiasma frequency or altered pattern of chiasmata
4. Restricted number of pairing initiation sites
5. Differential pairing in chromosome arms leading to chiasma interference or localization
6. Correction of multivalent SCs to bivalent pairing
7. Negative chiasma interference between opposite arms of quadrivalents and positive interference between adjacent arms

1. Genetic Control of Pairing

Genetic control of polyploid pairing is an idea which has been discussed by a number of workers (e.g., Avivi,[96] Charpentier,[97] Jackson[82]) to explain the pairing behavior of certain autotetraploids. Avivi[96] studied metaphase I pairing in autotetraploids derived from two lines of diploid *Triticum longissum*: one (7214), a promoter of homoeologous pairing in hybrids with common wheat and the other (7011), a suppressor of homoeologous pairing. The two diploid lines also differed in chiasma frequency, the 7214 line having three times as many interstitial chiasmata as the 7011 line. At metaphase I, the derived autotetraploids did not differ in chiasma frequency, but the 7214 tetraploid again had more interstial chiasmata. The frequency of quadrivalents in the 7214 tetraploid (65%) fitted the expectation of random pairing models,[36,83,85] whereas the quadrivalent frequency in the 7011 tetraploid (49%) was significantly lower. Avivi[96] concluded that the homoeologous pairing-suppressor genotype modifies homologous pairing in the autotetraploid to induce bivalent pairing, and suggested that this was via control of premeiotic spatial distribution of homologous chromosomes, that is, in the 7214 tetraploid, all four homologs of a tetrasome are associated premeiotically and have equal chances to pair, whereas, in the 7011 tetraploid, the homologs tend to be arranged in twos and hence pair more often as bivalents.

Charpentier et al.[97] compared metaphase I pairing in colchicine-induced autotetraploids of *Agropyron elongatum* with that in naturally occurring autotetraploids. They found that the induced autotetraploids had two to three multivalents per cell (significantly fewer than the expection from random pairing) and a reduced chiasma frequency, while the natural autotetraploids had close to 100% bivalent pairing. In crosses between the natural and induced tetraploids, the hybrids had quadrivalent frequencies similar to the induced line, suggesting that the weaker bivalent inducing system in the diploid progenitor was dominant to the genetic control of bivalent pairing in the natural tetraploid. Other synthetic autotetraploids with lower than expected quadrivalent frequencies include *Festuca pratensis*,[98] *A gropyron cristatum*,[99] and *Avena strigosa*.[100]

Avivi[101] used colchicine treatment immediately prior to meiosis to induce tetraploid microsporocytes in *T. longissum* and *T. speltoides*. At metaphase I, these tetraploid cells had almost exclusive bivalent pairing. Since these tetraploid meiocytes were formed by spindle failure in a premeiotic mitosis, the author concluded that this event occurred in the last premeiotic mitosis, so that sister chromatids remained closely associated at the beginning of prophase I and SCs formed preferentially between them. Avivi[101] concluded that the colchicine effect produced a phenocopy of the diploidizing genes and suggested that spatial order of chromosomes was important in determining pachytene pairing configurations.

Several other studies confirm that duplication of chromosomes immediately prior to pairing results in this type of exclusive bivalent pairing. Lelley et al.[102] found almost no multivalent formation at metaphase I in autotetraploid cells occurring spontaneously in rye. An exceptional case in autotetraploid *Allium odorum* was described by Häkansson and Levan.[103] Meiosis in the pollen mother cells was normal, with a mean of 43% quadrivalent formation at metaphase I. Chiasma localization near centromeres, as in *A. porrum*,[64] probably was the main reason for this low quadrivalent frequency.[103] In female meiocytes, the embryo mother cells, it appeared that an endomitotic division occurred just prior to prophase I, so that each chromosome was represented by two sister chromosomes lying side-by-side.[103] During prophase I, the two sister chromosomes paired so that 32 (the somatic number) bivalents formed. Chiasma frequency at metaphase I was a high four per bivalent. The remainder of female meoisis was normal, resulting in meiotic products with the somatic number. Seed formation and fertility was high and the tetraploid chromosome number continuity was maintained by an apomictic development of the embryo sac following incomplete fertilization by 16 chromosome pollen.[103] This is not unlike the situation during oogenesis in the unbalanced polyploid thelyokus parthenogenetic stick insect *Carausius morosus*,[104] where an early prophase I duplication of chromosomes is followed by a tetrapachytene with a somatic number of synaptonemal complexes formed by pairing of sister chromosomes (autobivalents).

However, these exceptional cases do not seem to be particularly relevant to the situation prevailing in most autotetraploids. Ultrastructural evidence generally does not support premeiotic chromosome association as a controlling factor in bivalent pairing;[105] hence, factors in addition to prophase I pairing behavior must be involved in determining the metaphase I chromosome configurations. In autotetraploid *A. vineale*, Loidl[74] found associations of four axial elements at early zygotene and high quadrivalent synaptonemal complex frequency at pachytene, but only 22% quadrivalent frequency occurred at metaphase I. Gillies et al.[48] and Rasmussen[42] found no reduction in early prophase tetrasome pairing in organisms with reduced metaphase I quadrivalent frequencies.

2. Pairing Preferences

Where autotetraploids are derived from heterozygous or outbreeding diploids, it is possible within a tetrasome that the four homologs may be present as two slightly different identical pairs of chromosomes. Pairing preferences between these may lead to preferential bivalent formation at the expense of quadrivalents. Orellana and co-workers[106-110] have examined this question with light microscopy of diakinesis/metaphase I, using C-band heteromorphic markers in rye and grasshoppers. In spontaneous tetraploid spermatocytes resulting from early stem cell chromosome doubling in the grasshopper *Eucharyhippus pulvinatus gallicus*, the three smallest chromosomes showed no quadrivalent formation and a significant excess of bivalents formed by pairing between identical chromosomes.[106] Quadrivalent frequency in the three largest chromosomes was 52%, but was only 10% in chromosomes 4 and 5. A similar situation was observed in teraploid cells from the grasshopper *Eyprepocnemis plorans*.[107]

In both these cases, with many of the chromosomes being telocentrics, it appears that pairing initiation sites are limiting, particularly in smaller chromosomes, but, in addition, there are pairing preferences between homologs.

Santos et al.[107] also found pairing preferences in autotetraploid rye which had C-band heteromorphic markers in chromosomes 1R[L] and 2R[S]. In this case, there was an excess of homologous over identical pairing in bivalents and quadrivalents in one or both chromosomes in some plants, while random pairing was found in other plants. Subsequently, Naranjo and Orellana,[108] in a more complete analysis of metaphase I pairing in the same chromosomes, found that 2R showed a significant excess of homomorphic (identical) pairing in both bivalents and quadrivalents in one plant, whereas another plant showed an excess of heteromorphic pairing in the bivalents. There was no evidence of preferential pairing in chromosome 1. The frequencies of quadrivalents differed from expectations for random pairing, but the authors considered that the limited degree of preferential pairing detected was not sufficient to explain the observed excess of metaphase I bivalent pairing.[108] They suggested that spatial separation of homologs or synaptonemal complex correction might be the cause of the increase in bivalent pairing. They also found differences between bivalents and quadrivalents in chiasma formation for most chromosome arms, which they attributed to partner switches interfering with chiasma formation and distribution in quadrivalents. Finally, they demonstrated that alternate orientation occurred in the majority of both ring and chain quadrivalents of chromosomes 1R, 2R, and 5R.[108]

Orellana and Santos[109] studied pairing preferences in both arms of chromosome 1R of autotetraploid rye marked with C-band heterozygosities. Overall pairing preferences were found for both bivalents and quadrivalents, the total result being 17 identical, 3 homologous, and 17 random associations. The fact that bivalents showed a higher tendency for identical pairing, and that bivalents and quadrivalents showed different pairing preferences suggested to these authors that partner switches in multivalents might reduce crossover frequency in pachytene quadrivalents, resulting at metaphase I in bivalents with identical long arms paired.[109] Other factors which were mentioned as possibly important were differences in chromosome condensation, number of interstitial pairing initiations, SC effects on crossing over, and correction of synaptonemal complex pairing.[109] Benavente and Orellana[110] found an excess of identical pairing in the long arm of 1R in tetraploid rye. The results from Orellana's group[106-110] lend some weight to the idea that pairing preferences occur among the nonidentical members of a tetrasome, but they also show that in many cases pairing preferences are not the major factor determining the reduction in quadrivalent frequency below theoretical expectations. Sybenga,[111] in a reanalysis of the data of Lentz et al.[92] from *Dactylis*, concluded that preferential pairing could be ruled out as a cause of low metaphase I quadrivalent frequency.

In tetrahaploids (isogenic tetraploids[112]) derived from haploids, all four chromosomes of each tetrasome set should be identical. Hence, such factors as preferential pairing (genetic or structural) should be eliminated. Callow et al.[112] found that the number of quadrivalents in tetrahaploid barley was not significantly different from the two thirds frequency expected if identical chromosomes paired randomly with one initiation per arm; 64% of anaphase I segregations were numerically balanced in this material. In contrast, Schlegel et al.[113] found only 41% quadrivalents in tetrahaploid rye, lower than the frequency of 47% in a related autotetraploid. A reduced chiasma frequency in the tetrahaploid could explain this difference, but the low quadrivalent frequency of the tetrahaploid still implies that factors other than preferential pairing must be capable of restricting the metaphase I quadrivalent frequency.

3. Chiasma Frequency Effects

Low chiasma frequency may be a factor limiting the formation of metaphase I quadrivalents in some organisms,[74] but increasing chiasma frequency does not always lead to increased quadrivalent formation at metaphase I. In an autotetraploid maize which had 85% quadrivalent formation,[114] the addition of abnormal chromosome 10 (K10) increased the chiasma frequency (and proportion of ring quadrivalents) without leading to any change in total quadrivalent frequency. Presumably, the existing high numbers of pairing initiations and crossovers among the four chromosomes of a tetrasome were already sufficient to ensure that quadrivalents persisted in most cases. It is debatable whether a higher ring quadrivalent frequency might lead to more balanced anaphase I segregation and improved fertility. Macefield and Evans[115] found that B (accessory) chromosomes increase the frequency of bivalents and decrease multivalents in autotetraploid *Lolium perenne*, apparently by altering the distribution, but not the frequency, of chiasmata. B chromosomes can thus influence both homologous and homoeologous pairing (see below) in a manner similar to genetic effects on the normal (A set) chromosomes.[96] In the genus *Senecio*, which has relatively low chiasma frequencies, triploid, tetraploid, and pentaploid hybrids between species at the diploid, tetraploid, and hexaploid levels have an increased chiasma frequency relative to the even ploidy species,[116] even though there is significant univalent formation in the hybrids and almost perfect pairing in the pure species. Ingram and Noltie[116] suggest that increased chiasma frequency in the hybrids compensates for asynapsis, possibly due to the increased availability in hybrids of precursors normally limiting crossing over in this genus and by crossing over at pairing-partner switches. This example shows similarities to the situation in the lily "Black Beauty"[117] (see below).

4. Effect of Number of Pairing Sites

Watanabe has studied a number of *Chrysanthemum* taxa ranging from tetraploid to decaploid,[181-121] as well as colchicine-induced tetraploid and dodecaploid lines.[122] He found almost exclusive bivalent formation at metaphase I in tetraploid, hexaploid, octaploid, and decaploid species, and very high bivalent frequencies in even ploidy hybrids between them, but odd ploidy hybrids had many more multivalents,[119] suggesting that random pairing of multiple homologs was possible. Thus, preferential pairing was not a likely explanation for bivalent pairing in the even ploidy taxa. Watanabe[119] favored an explanation based on genetic suppression of pairing sites in one arm of the chromosomes, so that bivalent formation occurred exclusively. Release of this suppression in triploid hybrids would allow the trivalent pairing observed.

In colchicine-induced tetraploids and dodecaploid *Chrysanthemum*, Watanabe[122] found metaphase I bivalent pairing frequencies of 84 and 96%, respectively, the same pairing behavior seen in the naturally occurring polyploids. Watanabe concluded that the suppressor system already existed, even at the diploid level, prior to chromosome doubling.[122] He suggested that the evolution of the multivalent suppressor system in polyploids involved the reduction of pairing initiations to a single site per chromosome, accompanied by a differentiation of recognition sites so that, ultimately, homoeologs are produced which do not normally pair.[122] Analyses such as those of Hamey et al.[39] indicate that data from other species support more than one pairing initiation per chromosome. While no zygotene/pachytene data exist for *Chrysanthemum*, the results of Loidl and Jones,[34] Loidl,[74,75] and Gillies et al.[48] indicate that the number of pairing initiation sites in other plant polyploids is many more than one per chromosome arm. Comparison of the results from triploid and tetraploid *Allium* species with chromosomes of similar length[34,74] reveals that both have multiple pairing initiations at zygotene, producing high multivalent frequencies at pachytene, but that the metaphase I multivalent frequency of the tetraploid is only about one fifth that of the triploid. Hence, factors other than pairing initiations (e.g., crossover effects) may be important in determining the type of metaphase I pairing observed in autopolyploids.

E. Allopolyploidization of Autopolyploids

Naturally occurring and artificially produced autotetraploids often incur a penalty of reduced fertility because of problems with the orientation and disjunction of multivalents and univalents at metaphase I. In most instances, the sterility arises from the disjunction of the occasional trivalents and univalents.[123] Long-established naturally occurring autopolyploids which have almost exclusively bivalent pairing at metaphase I (e.g., *Avena barbata*,[100] *Physaria vitulifera*,[124] *Poa annua*,[125] and several *Rhynchosinapsis* species[126] may be considered to have been diploidized. As part of efforts to utilize autotetraploid plant species, breeders have often attempted to induce allopolyploid-like behavior in autotetraploids.[127,128] This may take the form of causing chromosome differentiation with rearrangements,[128] utilizing genetically determined pairing preferences,[128] or a combination of the two. Doyle[129] has used this approach in autotetraploid maize, with X-rays and chemical mutagenesis-induced differentiation of genomes resulting in a reduction in quadrivalent frequency from 81 to 73% over 10 generations. Gilles and Randolf[130] had found a similar decrease in quadrivalent frequency in autotraploid maize over 10 years, with a corresponding increase in bivalents. In this case, the change was the result of selection for more vigorous and fertile plants.

Studies of the relationship of fertility and chromosome pairing have not always yielded expected results. Mastenbroek et al.[131] reported results from autotetraploid maize lines mass selected for seed set, yield, and plant and ear height for up to 22 generations. While seed set went up from 57.5% to 80 to 85% over the period of the experiment, quadrivalent and bivalent frequencies at diakinesis were not significantly changed. Thus, improved fertility resulted not from changes in chromosome pairing, but presumably from selection for balanced gametes, i.e., postmetaphase I behavior. In autotetraploid rye[132] and *Lolium*,[133] fertility was positively correlated with quadrivalent frequency, the opposite of an expected diploidizing effect. This was largely a result of increasing and relocating chiasmata. Hossain[134] also reported that 20 years of random mating after the establishment of an outbred rye tetraploid line appeared to increase quadrivalent frequency. However, compared with the inbred tetraploid lines reported by Hazarika and Rees,[132] the quadrivalent frequency was lower and the chiasma frequency was higher in the outbred, randomly mating line. Hossain[134] concluded that the outbreeding tetraploid rye had a more disomic association of chromosomes at zygotene as a result of accumulated chromosome differences, whereas the inbred rye has a tetrasomic association of chromosomes at zygotene. In addition, Hossain and Moore[135] found that four generations of selection for high seed set and regular tetrads in the outbred rye resulted in a significant reduction in metaphase I quadrivalent frequency, and a corresponding significant increase in bivalents. Chiasma frequency was positively correlated with quadrivalent frequency.[136]

IV. ALLOPOLYPLOIDS

A. Homoeology

Allopolyploids usually arise by chromosome doubling following hydridization.[20] Hybridization of two species requires some degree of relationship, so that partial homology might be expected between chromosomes of the two parental species (see Chapter 4). Thus, both homology and homoeology are possible in allopolyploids and it is reasonable to expect variability in the pairing behavior of allopolyploids. An example of this situation is the artificial allotetraploid *Lycopersicon esculentum — Solanum lycopersicoides*.[137] The diploid F1 exhibited a high degree of pachytene pairing and normal SCS in sections,[138] with an average of one chiasma per metaphase I bivalent.[137] However, the allotetraploid had almost complete preferential pairing at pachytene, with only occasional quadrivalents and 95% bivalents at metaphase I. The chiasma frequency in the allotetraploid was as high or higher than in the parental species.[137] Thus, in this alloploid there existed an ability to discriminate exact homologs from homoeologs during pairing.

As studies of the control of pairing become more extensive, and as we begin to accumulate data on zygotene/pachytene behavior in allopolyploids, it begins to appear that strict allopolyploidy with exclusively autosyndetic pairing is the exception. By far the more common situation is some form of segmental allopolyploidy in which both autosyndesis and allosyndesis occur, but where the pairing may be under genetic control. As in autopolyploids, irregularities in pairing or segregation in allopolyploids can lead to unbalanced gametes and infertility. Thus, selection for regular pairing and segregation in ancient allopolyploids might be expected to have favored the development of control systems to regularize pairing. This control may act on the choice of lateral element pairing partner at zygotene (or earlier), on the rate and position of synaptonemal complex initiations, on the stability of zygotene/pachytene synaptonemal complexes, or on the number and position of crossover events.

The majority of recent work has been carried out on the pairing behavior of polyploid cereals and grasses, particularly wheat and its relatives. These species include many which obviously can be classified as segmental allopolyploids. Outside this group, the few recent studies are generally confined to light microscope studies of metaphase I pairing (e.g., *Brassica* species[139]). Two recent studies of prophase I which utilize SC spreading techniques illustrate the pachytene pairing behavior of strict autosyndesis[31] and classical segmental allopolyploidy.[140] The allotetraploid *Allium senescens* (subsp. *montanum* = *A. monatanum* has only bivalents at diplotene and metaphase I. From mid-zygotene onwards, Loidl[31] found only bivalent synaptonemal complex formation, and complete bivalent pairing at pachytene. Apparently, pairing is restricted to homologues from the very first and pachytene pairing is strictly autosyndetic. The tetraploid *Paeonia officinalis* studies by Schwarzacher-Robinson[140] is reported to be an allotetraploid with few karyotypic differences to distinguish the parental genomes. Pachytene electron microscopic spreads revealed that pairing was never complete. Two to four axial elements often ran close and parallel to each other and paired into synaptonemal complexes, with partner switches forming trivalents and quadrivalents. Unpaired axial elements were common and completely paired bivalents, rare. Since the species had univalents, bivalents, trivalents, and quadrivalents at metaphase I, Schwarzacher-Robinson[140] concluded there was sufficient homology between the genomes to allow some pairing and crossing over, but pairing deficiency resulted in a high number of univalents at metaphase I. It would appear that *P. officinalis* is a true segmental allotetraploid with no genetic controls on pairing. Lateral element thickenings observed may be a consequence of pairing difficulties experienced by chromosomes from genetically different genomes (see Reference 28).

B. Wheat Genomes and the *Ph* Gene

It is generally accepted that common bread wheat (*Triticum aestivum*) is an allohexaploid derived from three diploid species whose genomes have been designated AA, BB, and DD.[141,142] Sears[143] established that these three genomes are related and composed of homoeologous chromosomes. However, while the homoeologs from the three genomes have the potential to pair and undergo crossing over, only bivalent pairing between homologs is usually seen at metaphase I in *T. aestivum*. This autosyndetic behavior is under complex genetic control, with the principal control exercised by a gene locus designated *Ph* (for Pairing homoeologous[144]) located on the long arm of chromosome 5B.[145,146]

In the absence of the *Ph* locus (as in nullisomic 5B — Riley[147]), wheat exhibits allosyndetic pairing behavior and may form multivalents of 3, 4, 5, or 6 chromosomes at metaphase I. The *Ph* gene effect is manifested in the hemizygous condition in euhaploid *T. aestivum* (2n = 3x = 21) which had a mean of 1.15 bivalents and 18.67 univalents per metaphase I cell, while the nulli 5B haploid (2n = 20) had a mean of only 7.50 univalents, 3.83 bivalents, 1.50 trivalents, and occasional higher number associations.[147] Absence of the 5B locus in hybrids between *T. aestivum* and other species had the effect of increasing chiasma frequency[148,149] (see also below). At metaphase I in plants tri-isosomic for $5B^L$ (i.e., with six doses of the *Ph* gene), some homoeologous pairing occurs, but total pairing (chiasma frequency) is reduced.[150] In addition, with higher doses of *Ph*, the interlocking of bivalents becomes much more common than in normal wheat.[151]

A number of mutations of *Ph* loci have been obtained[144,152,153] and two were mapped to the long arm of chromosome 5B.[141,154] Detection and monitoring of the effect of these *Ph* mutants has often been by analysis of metaphase I chromosome associations in hybrids between *T. aestivum ph* and other species of *Triticum, Aegilops*, and *Secale*. Sears[152] showed that the mutant he produced with X-rays (*ph1b*) increased (homoeologous) metaphase I pairing in hybrids with *T. kotschyi* (2n = 4x = 28 $C^UC^US^VS^V$) from 1.03 bivalents per cell to 7.65 bivalents, 2.70 trivalents, 0.75 quadrivalents, and 0.40 pentavalents per cell. The chiasma frequency in the *ph1b* hybrids was about 50% greater than in nulli 5B hybrids. This was interpreted as a consequence of the presence of pairing promoters on $5B^S$ in *ph1b* stocks and their absence in nulli 5B.[152] Another mutant was isolated from tetraploid wheat (*T. turgidum* var. *durum*, 2n = 4x = 28 AABB), which increased metaphase I chromosome associations in hybrids with *T. kotschyi* and *A. cylindrica* (2n = 4x = 28 CCDD).[153] This mutant has been designated *ph1c* and is thought to be caused by a deletion of the *Ph* locus on $5B^L$.[154,155]

Although the *Ph1* locus exerts the major control on homoeologous pairing in wheat, other controls have also been discovered. Pairing suppressors also reside on chromosomes $3D^S$ (*Ph2*[156,157]), $3A^S$, $3B^S$, $2D^L$,[158] and possibly 7A.[159] Pairing promoters have been found on chromosomes $5A^S$, $5B^S$, $5D^S$, $5A^L$, $5D^L$, $3A^L$, $3B^L$, $3D^L$, $2A^S$, $2B^S$, and $2D^S$.[150,158,160] The latter controls appear to work in opposition to the *Ph* loci and their effects can be monitored by use of telosomic aneuploids to show that extra doses of these chromosome arms increase homoeologous pairing, whereas a deficiency for them decreases homologous pairing.[158,160,161]

C. Homoeologous Pairing in Other Cereals

Genetic control of homoeologous pairing in allopolyploids of other cereals, often closely related to wheat, is not uncommon. An incomplete list of species shown to carry genes for regulation of homoeologous pairing includes *Agropyron junceum* (6x),[162] *A. elongatum* (2x),[163] *T.(Aegilops) speltoides* (2x),[164] *A. ventricosa* (4x),[165-167] *Hordeum* spp. (6x and 4x),[168] *Avena sativa* (6x), and *A. longiglumis* (2x).[169,170] In most cases, the nature and number of genes involved has not been elucidated.

More intensive studies in the *Festuca-Lolium* group have identified both genetic and cytological factors involved in the control of polyploid pairing. In tetraploid hybrids between hexaploid *F. arundinaceae* and diploid *L. perenne*,[171,172] three types of control were recognized: diploid species genes, B (accessory) chromosome effects, and genes from natural polyploids. Pairing in the hexaploid *F. arundinaceae* seems to involve genetic control of bivalent homolog pairing within each of the three closely related genomes.[173,174] Homoeologous pairing control occurred in one monosomic *F. arundinaceae* line, and a euhaploid had up to seven bivalents, suggesting that polyploid genetic control was ineffective in the hemizygous condition.[174]

The polyploid genetic system in hexaploid *Festuca* interacts with the genetic and B chromosome controls from diploid *L. perenne*.[171] Evans and Macefield[175] have shown that the presence of 4 B chromosomes in tetraploid *L. temulentum* × *L. perenne* increased metaphase I pairing frequency from 6.41 to 12.8 bivalents per cell (see Chapter 4, Figure 11). That is, B chromosomes suppress homoeologous associations. Aung and Evans[176] found that similar controls in this allotetraploid can be exerted by genetic factors on the normal chromosomes from both species. Analysis of marker segregation confirmed that the suppressors restricted crossing over to homologs.[177]

As discussed by Jenkins in Chapter 4, ultrastructural analyses of zygotene and pachytene pairing have been carried out on *L. perenne* × *L. temulentum* allopolyploids. In both triploid and tetraploid plants, it was found that B chromosomes and pairing genes which suppress homoeologous associations at metaphase I appear to have little effect on the zygotene pairing behavior.[178,179] Homoeologous chromosome associations and multivalent SCs were formed at zygotene in the presence or absence of the controlling factors.[179] By pachytene in the triploid with B chromosomes, the homoeologous pairing was resolved into homologous bivalents plus nonhomologous univalent associations.[178] In the tetraploid with both B chromosomes and pairing genes, pachytene pairing was largely as bivalents, but some multivalents persisted. Since metaphase I pairing was strictly homologous, the controls must act to promote resolution of homoeologous to homologous pairing, but must also restrict crossing over to homologously paired regions,[180] possibly in a similar manner to the *Ph* gene in wheat, as discussed further below.

The chromosomes of diploid common rye *Secale cereale* have limited homology with the wheat homoeologous sets.[161,181,182] In hybrids and allopolyploids, there is usually only occasional metaphase I pairing between wheat and rye homoeologs, possibly due to differences in metabolic activity or cell cycle behavior.[183] The majority of wheat-rye homoeologous pairing in one study involved the chromosome 1 (NOR bearing) group.[182] The absence of wheat chromosome 5B or the *Ph* gene from wheat X rye allopolyploids increased the metaphase I pairing between wheat homoeologs, but caused only a small increase in pairing between wheat and rye homoeologs.[161,184,185] There is good genetic evidence that crossing over occurs between wheat and rye homoeologues in the absence of *Ph*. Koebner and Shepherd[186] found up to 1.44% recombinants between rye $1R^L$ and wheat $1D^L$ in a *phlb phlb* $1R^L/1D^S$ translocation substitution line and confirmed that recombinants between IR^S and $1D^S$, and $1R^S$ and $1B^S$, could also be obtained.[187,188]

Models of pairing in triploid and tetraploid hybrids have been devised[189,190] which allow the estimation of affinity between genomes in allopolyploids. The models rely on metaphase I pairing data, including the frequency of open and closed bivalents and quadrivalents. Kimber and Alonso[190] used their model to show that, in the tetraploid hybrid between hexaploid wheat and diploid rye, the three wheat genomes were relatively much more homologous with each other than any one was with rye. The wheat A and D genomes showed some preferential affinity for each other. Variation in dosage of the *Ph* pairing control gene did not affect the measurement of pairing affinities.

However, in wheat X rye allopolyploids, it appears that the presence of rye chromosomes increases the metaphase I homoeologous pairing between wheat chromosomes, perhaps by suppressing the wheat *Ph* gene effect.[191,192] Of six rye species tested, *S. montanum* had the strongest effect, causing up to 2.4% wheat-rye pairing.[193] Within 39 *S. cereale* lines crossed with Chinese Spring wheat, there was evidence for both major and minor rye genetic systems which affected homoeologous pairing in wheat.[194]

Only one study has attempted to investigate the pachytene behavior of wheat X rye allopolyploids. Abirached-Darmency et al.[195] carried out a limited serial section analysis of synaptonemal complex formation at pachytene in diploid (DR), triploid (ABR), and tetraploid (ABDR) wheat X rye hybrids. In each case, the metaphase I pairing resulted in an average of less than one bivalent per cell. Although formation in the diploid was extremely limited, one triploid had 25% of the lateral elements paired as SCs and one tetraploid had 35% paired lateral elements.[195] Reconstructions in the triploid revealed a trivalent and quadrivalent, which was considered evidence of wheat-rye pairing. In addition, there were lateral element thickenings in some of the SCs in the triploid, and in most of the few SCs present in the diploid.[195] These could be the consequence of wheat-rye pairing forming SCs between partially nonhomologous regions of homoeologs[195] (see Reference 28). It was thought that most of the synaptonemal complexes in the ABDR tetraploid probably involved wheat homoeologs (particularly A and D genomes) since no such thickenings were present in SCs in the tetraploid.[195] These results add to the evidence that homoeologous synapsis can occur at zygotene in wheat, but it does not necessarily result in homoeologous associations at metaphase I. The presence of chromosome 5B in the triploid and tetraploid hybrids was presumably instrumental in preventing homoeologous crossovers, although recombination nodules were observed in the tetraploid ABDR.[195] Nulli 5B ABDR hybrids with up to 70% pairing at late zygotene and 7 metaphase I chiasmata have recently been reported.[229]

D. Models of *Ph* Gene Action — Light-Microscopic Data

There has been considerable speculation and debate about the mode of action of the *Ph* gene in regulating homoeologous pairing in wheat. Riley[149] put forward a hypothesis based on the contention that homologous and homoeologous pairing were at least partially separated during prophase I, so that in the first or attraction stage both homologs and homoeologs could associate imprecisely, while the second stage only precise homologous pairing was possible. Riley[149] proposed that in hexaploid wheat the *Ph* gene product in some way curtailed the first stage so that only second-stage pairing between homologs would occur, leading to strict bivalent pairing at metaphase I. In nullisomic 5B or *Ph* mutants, the absence of the *Ph* product meant that the attraction phase was prolonged and both homologous and homoeologous pairing occurred. Moreover, additional doses of the *Ph* gene would terminate the attraction phase even earlier so that pairing would be reduced and homologs and homoeologs would remain tangled and interlocked and fail to pair completely, as found in di-isosomic and tri-isosomic 5B^L.

Some support for this theory seemed to come from the results of Bennett and Kaltsikes,[183] who found that meiotic and prophase I times in diploid rye, tetraploid and hexaploid wheat, and hexaploid triticale were all different. This could result in a time limit for pairing in hybrids, and affect meiotic stability. However, subsequent studies in hexaploid wheat aneuploids showed that the duration of meiosis was not correlated with the amount of homoeologous pairing.[196] An effect of nullisomic 5B was shown to be due to 5B^S. Thus, if the *Ph* gene affected pairing times, it did not alter total meiotic duration.

Feldman and co-workers[197-202] have proposed an explanation of the *Ph* gene action which is based on the somatic association of chromosomes prior to meiotic pairing. According to this hypothesis,[197] the positions of chromosomes in premeiotic cells were instrumental in deciding their choice of pairing partner during meiosis. Homologs and homoeologs tend to be more closely associated than nonhomologs in premeiotic nuclei and, consequently, a low level of nonhomologous pairing is usually observed. Feldman[197] suggested that the *Ph* gene product in normal hexaploid wheat acted on the spatial distribution of chromosomes, suppressing homoeologous associations more than homologous ones, such that meiotic pairing was largely between homologs and bivalents formed at metaphase I. In the nullisomic 5B or *ph* mutant, absence of the *Ph* product allowed somatic association of both homologs and homoeologs and meiotic pairing could then include both. The hypothesis could also accommodate the effect of multiple *Ph* doses since Feldman[197] proposed that these would further suppress somatic association so that meiotic pairing became more random, resulting in the observed reduction of chiasma frequency, multivalents, heteromorphic bivalents, univalents, and interlocking.[150] Subsequent reports from Feldman and co-workers have produced evidence in support of the model and attempted to determine its cellular and biochemical basis.

Avivi and Feldman[198] presented evidence that the *Ph* product acts through the premeiotic spindle system to determine chromosome positioning. They suggested that in nullisomic 5B^L the centromeres of homologs and homoeologs converge, while, with two doses of 5B^L, the spindle separates homologous from homoeologous centromeres, but not homologous from homologous. With six doses of 5B^L, all centromeres (homologous and homoeologous) are separated premeiotically and subsequent meiotic pairing is random. In support of this theory, analyses of root tip mitotic chromosomes of hexaploid wheat with one or two doses of the *Ph* gene have shown that the chromosomes of each genome are relatively separated from those of the other genomes, and that homologs lie closer together than homoeologs.[199,200] However, with zero doses of *Ph*, there was no difference between the arrangement of homologs and homoeologs.[200] Analyses of metaphase I bivalent disposition in tetraploid and hexaploid wheat also revealed that bivalents from each genome were clustered and, in comparisons between genomes, homoeologous bivalents were closer than nonhomoeologs.[201,202] Yacobi et al.[151] have confirmed the dosage effect of *Ph* genes on metaphase I interlocking, but also discovered that with only zero or one dose of *Ph* there was an increase in interlocking, although possibly only homoeologous interlocks. 5B^S appeared to have a suppressing effect on interlocking.

The idea that the *Ph* gene product acts on the spindle has part of its origin in observations of the effects of colchicine on meiotic chromosome pairing. Dover and Riley[203] found that colchicine applied during early premeiotic interphase in wheat induced univalent formation (i.e., reduced chiasma frequency) at metaphase I. Together with other studies, this led to the suggestion by Avivi and others that the mode of action of the *Ph* gene product might be via an effect on the spindle. In recent studies, they have shown that the spindles in root tip cells of wheat are more colchicine sensitive in *phlb/phlb* stocks than in *Ph/Ph* stocks.[204] Low doses of griseofulvin, a drug which binds to tubulin or microtubule associated proteins (MAPs) and interferes with microtubule assembly and spindle polarity, showed no difference in effect in wheat with 0 to 4 doses of *Ph*.[205] At higher doses of griseofulvin, the 0 *Ph* wheat was more sensitive than the 2 or 4 *Ph* wheat, suggesting that the *Ph* gene product could be tubulin or a MAP which stabilizes microtubules. This was supported by the fact that iso-propyl-*N*-phenylcarbamate (IPC), which appears to act on cytoskeletal elements and micro-tubule organizing centers rather than on microtubules per se, did not act differentially in wheat with 0 to 4 *Ph* doses.[205] Ceoloni and Feldman[206] have recently shown that root tips of plants with different doses of *Ph2* (3D^S) have different colchicine sensitivity. Nullisomic 3D^S and two different mutant alleles, *ph2a* and *ph2b*, reduced sensitivity to colchicine, compared with wild type. This is the opposite of the *Ph1* effect, leading to the conclusion that both genes act on microtubule-associated cell functions, perhaps affecting conformational changes with opposing effects on the spindle. but in both cases increasing homoeologous pairing.[206]

An alternative suggestion is that colchicine acts at or near the nuclear membrane by affecting one or more membrane-associated proteins. In lily, colchicine applied from pre-meiotic to mid-zygotene reduced chiasma formation[207,208] and this effect appeared to result from the failure of lateral elements to pair into SCs. Hotta and Shepard[209] found that, in lily meiotic cells, the principal biochemical effect of colchicine was a reduction of nuclear membrane-associated DNA binding protein in early prophase I. This protein has a role in DNA strand matching during zygotene and pachytene. No effect was found after the protein synthesis was complete at pachytene. They also found another colchicine-binding protein in the nuclear membrane, distinct from the DNA-binding protein.[209] These membrane proteins provide an alternative basis in lily for the action of colchicine, besides interaction with the spindle. Ultrastructural studies in wheat and other cereals[210,211] revealed the presence during premeiotic, leptotene, and zyogtene meiotic stages of intranuclear fibrillar material (FM), which was sometimes associated with the nuclear envelope and chromosome telomeres. FM peaked in abundance before pairing was complete and was absent from pachytene onwards. This suggested a possible function in attachment and movement of chromosome telomeres at the nuclear envelope during pairing.[211] Since the total amount of fibrillar material formed was not affected by colchicine treatment, Bennett and Smith[212] concluded it was not tubulin. However, colchicine applied premeiotically did retard FM formation during the period in which colchicine depresses chiasma formation.[203] Bennett and Smith[212] commented on the coincidence of a nuclear envelope-associated colchicine-binding protein, fibrillar material attached to the inner nuclear envelope, and telomeric initiation of chromosome pairing at the nuclear envelope, implying that the three might be linked and possibly related to genetic effects on pairing. Loidl[228] has recently reported that colchicine applied to *Allium ursinum* microsporocytes during leptotene suppressed SC formation, possibly by interfering with chromosome alignment. In addition, at zygotene, when homologs were already aligned, colchicine prevented both SC completion (central element formation), and length-wise extension of pairing initiation sites.[228] Although colchicine may be considered to produce phenocopies of the genetic effects,[212] there is some doubt about whether it acts in the same way as *Ph* and other genetic controls.

On the basis of mathematical analyses of metaphase I chromosome associations in hybrids, aneuploids, and colchicine-treated wheat, Driscoll et al.[213] concluded that homologs and homoeologs in hexaploid wheat associate presynaptically in groups of six, with subsequent homologous associations being more efficient. Genes on chromosomes of homoeologous groups 3 and 5 do not affect these associations, but alter the number of chiasmata and their distribution. Thus, the absence of 5B decreased the total number of chiasmata, but increased the proportion of homoeologous exchanges. It was estimated that, in two nullisomic 5B plants, an average of 5.3% of chiasmata involved homoeologs.[214] Driscoll[215] suggested that the mode of action of the 5B effect could be via production of an enzyme which determines whether pairing events lead to crossing over.[216] According to this theory, the *Ph* gene product normally prevents completion of homoeologous pairing so that crossing over between homoeologs does not occur. Removal of *Ph* gene product allows homoeologous pairing to proceed to the stage where crossing over can occur. This would seem to fit in with the results which show that effective pairing is necessary in order for crossing over enzymes to function.[117,217]

E. Models of *Ph* Gene Action — Electron-Microscopic Data

The last 7 years have seen a number of attempts to determine the cytological basis of homoeologous pairing control in hexaploid wheat by investigating SC formation at zygotene and pachytene in plants with various doses of active *Ph* genes. Reconstructions of SCs from serially sectioned hexaploid wheat have provided important results,[218-221] but the number of nuclei studied has been limited by the laborious nature of this method. Hobolth[218,219] produced complete three-dimensional reconstructions of SC formation in one late zygotene nucleus and one pachytene nucleus from wild-type hexaploid Chinese Spring wheat. The zygotene nucleus had 5 bivalents, 4 quadrivalents, 1 pentavalent, and 1 hexavalent, together with associations of acentric fragments and foldback paired univalents. The multivalents all had two or more partner switches, often two close together, and two interlocks were also identified. Hobolth[218] also reported pairing-partner switches in other incompletely reconstructed zygotene nuclei, indicating the presence of multivalents. In the pachytene nucleus, 21 bivalent SCs were found, all apparently homologously paired.[218] Thus, there was a change from some multivalent (presumably homoeologous) pairing at zygotene to strict bivalent (homologous) pairing at pachytene. Hobolth[218,219] concluded that, although most zygotene pairing in hexaploid wheat was between homologous chromosomes or telomeric segments of homologs, homoeologous pairing did occur regularly in interstitial regions. He speculated that crossing over did not occur until after the correction of homoeologous pairing to exclusive homologous pairing at pachytene. Recombination nodules were seen at both zygotene and pachytene, but their random distribution and high frequency indicated that not all would lead to crossovers. Hobolth[218,219] advanced the suggestion that the *Ph* gene controlled the time of crossing over relative to pairing correction; that is, in normal disomic 5B^L, crossing over occurred in pachytene after correction was complete, but in nullisomic 5B^L, crossing over occurred prior to the completion of correction, so that chiasmata were possible between regions still homoeologously paired. Extra doses of *Ph*, as in tri-isosomic 5B^L, delayed crossing over so long that chiasma frequency was reduced. Hobolth[218] suggested that *Ph* might code for a protein functional in meiotic prophase.

Another serial section study of synaptonemal complex formation in hexaploid wheat confirmed that multivalents were present at zygotene/pachytene[220] and centromeres were present in groups. Possible interlocks were seen, but chromosomes were aggregated in chromatin "knots" at zygotene, which prevented complete analysis. It was suggested that the "knots" might represent unresolved multivalents or interlocks. These knots disappeared by pachytene, and it was concluded that chromosome condensation changes occur in parallel with resolution of multivalents and interlocks.[220] Von Wettstein et al.[37] discussed these serial section data and reiterated the idea that *Ph* delays crossing over to allow pairing correction. They suggested that similar systems might operate in *Chrysanthemum,*[122] *Avena,*[170] and *Festuca.*[173]

The finding that homoeologous bivalents were significantly closer together at metaphase I than were nonhomoeologs,[201,202] presented as evidence for somatic association of homoeologs, could equally well be the result of homoeologous multivalent pairing at the preceding zygotene.

Yacobi et al.[151] criticized Hobolth's model on the grounds that if *Ph* delayed crossing over, then with higher *Ph* doses the frequency of interlocks should be reduced, which is opposite to their findings. Wischmann[221] investigated this problem, using both serial sectioning and surface spreading of synaptonemal complexes in tri-isosomic $5B^L$ plants which, at metaphase I, had both multivalents and interlocked bivalents and a mean of 17.9 chiasma per cell. One reconstructed nucleus had 25% of the lateral element length paired into synaptonemal complexes, including 58 of the 86 telomeres.[221] There was 1 association of nine lateral elements, 1 of five lateral elements, 6 of three lateral elements, and 12 of pairs of lateral elements, as well as nonhomologous and triple pairing, 6 knots, and at least 1 interlock. Thirty-three recombination nodules were identified,[221] about one third of the number seen by Hobolth[218] at late zygotene in normal wheat. All of the spread nuclei Wischmann[221] examined were incompletely paired, the maximum pairing observed being 37%. Pairing was terminally polarized and pairing switches were common. Multiple associations in the spread zygotene nuclei exceeded the multivalent frequency at metaphase I and foldback and triple pairing was also observed. Wischmann[221] concluded that the presence of six copies of $5B^L$ resulted in pairing being arrested at mid-zygotene and, although some correction or dissociation of nonhomologous pairing occurred, it was incomplete, so that crossing over led to nonhomologous associations at metaphase I. It was subsequently shown that four copies of $5B^L$ caused zygotene pairing arrest at a later stage than did six copies.[222] Evidence from both di-isosomic and tri-isosomic $5B^L$ supports the idea that resolution of interlocks is dependent on completion of SC formation.[221-223] Comparison of the effect of extra *Ph* copies with the reported effects of cold and colchicine led Wischmann[221] to conclude that, while these treatments acted to arrest pairing, extra doses of *Ph* reduced both pairing and crossing over.

Holm and co-workers have utilized the SC spreading technique to investigate pairing in wheats with a range of 5B dosages.[222,223] Holm[223] found that SC initiations at zygotene in normal hexaploid wheat occurred most commonly in distal regions and with a frequency per unit length similar to that reported in rye, *Tradescantia,* and lily. Pairing was predominantly into bivalents, but multivalent synaptonemal complexes were regular at mid-zygotene (Figure 6), although at a lower frequency than reported by Hobolth,[218] and they decreased to very occasional examples at pachytene, presumably as a consequence of correction to bivalent pairing.[223] Multivalents with two reciprocal pairing-partner switches over a short distance disappeared earlier than those with only a single switch. Interlocks were very frequent at mid-zygotene, but decreased in frequency to almost zero at pachytene, from which Holm[223] inferred that resolution of interlocks was related to pairing completion. Because there were still occasional multivalents at pachytene but none at metaphase I, Holm[223] suggested that crossing over in homoeologous and nonhomologous SCs must be suppressed in euploid wheat.

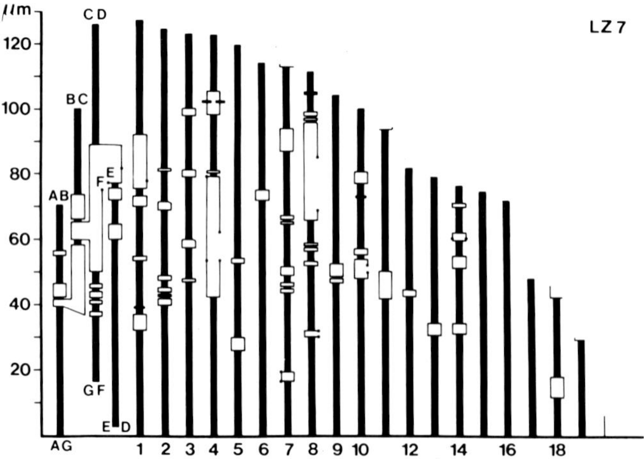

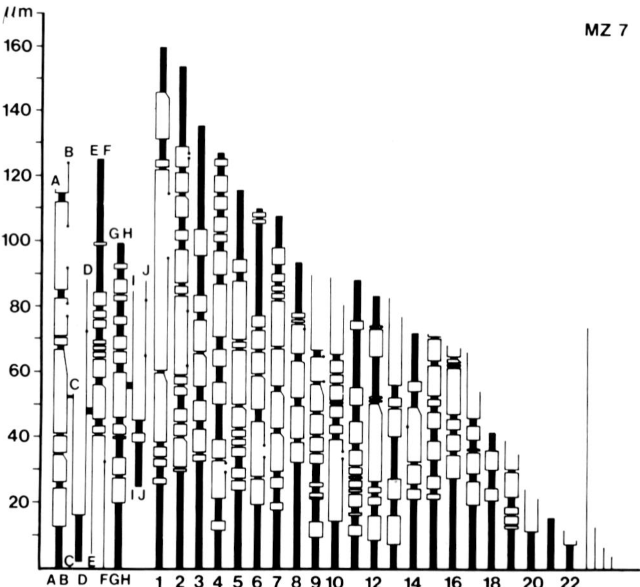

FIGURE 6.   Idiograms of spread synaptonemal complexes (heavy lines) and lateral elements (thin lines) from electron micrographs of a mid-zygotene nucleus (MZ7) and a late zygotene nucleus (LZ7) of Chinese Spring wheat. Arranged by decreasing length, with multiple associations (A-F,G-J in MZ7; A-G in LZ7) shown first and fragments last. Breaks in lateral elements are indicated by dots and NORs by cross bars. (From Holm, P. B., *Carlsberg Res. Commun.*, 51, 239, 1986. With permission.)

Monosomic 5B wheat, which usually has the maximum 20 bivalents at metaphase I, appears to have twice as many chromosomes associated in multivalents at zygotene as does euploid wheat.[222] Since correction of multivalent pairing occurred in the monosomic, but was not complete, some multivalents persisted at pachytene. Hence, although one copy of *Ph* allowed more homoeologous pairing, it was still sufficient to suppress homoeologous crossing over.[222] At zygotene in the absence of *Ph* (nullisomic 5B), up to 2.5 times as many chromosomes as in the euploid were present as multivalents and pairing was never more than 66% complete.[222] There appeared to be little correction of multivalent pairing in the nullisomic; hence, the absence of *Ph* has the combined effect at zygotene/pachytene of allowing more homoeologous pairing, arresting pairing before completion, and allowing some crossing over between homoeologs.[222] Holm et al.[222] concluded that the *Ph* gene product controls the stringency of zygotene homologous pairing, but also suppresses crossing over between homoeologs. In addition, absence of *Ph* or its presence in higher doses reduces SC formation, which has the side effect of preventing resolution of some interlocks.[222]

In haploids (trihaploids) derived from hexaploid wheat and hybrids of wheat with some other species, only homoeologous or nonhomologous pairing and crossing over is possible. As mentioned previously, homoeologous SC formation reached 35% in tetraploid (ABDR) hybrids between wheat and rye,[195] but the presence of one 5B chromosome appeared to suppress most homoeologous crossing over. In a spreading analysis of synaptonemal complexes in trihaploid Chinese Spring wheat, Holm et al.[222] found up to 40% pairing, whereas in trihaploids from the winter variety ''Kedong'', pairing was up to 90%. At early zygotene, the majority of lateral elements were paired in associations of three or more, but as pairing increased, the complexity of associations decreased, suggesting the correction of pairing occurred. Since there was a mean of only 0.17 bivalents per metaphase I cell, the single *Ph* gene obviously suppresses most homoeologous crossing over.[222] Trihaploid wheat nullisomic for 5B showed a threefold increase in the number of partner switches,[222] supporting the idea that *Ph* controls the stringency of zygotene pairing. Thus, the wheat nullisomic 5B trihaploid behaves at zygotene in the same manner as a strict haploid, such as that of barley.[224] However, the wheat nullisomic 5B haploid has bivalents, trivalents, and occasional higher associations at metaphase I,[147] indicating that crossing over occurs between nonhomologs, while crossing over is virtually absent in the barley haploid.[224]

Pentaploid hybrids between hexaploid *T. aestivum* and tetraploid *T. kotschyi* have five homoeologous genomes which show little homoeologous crossing over in the presence of the *Ph* gene, but have bivalents and multivalents at metaphase I if the *ph1b* mutant is present on chromosome 5B.[152] Studies of SC formation in spreads of these hybrids[225] revealed that both genotypes had extensive pairing at zygotene/pachytene, with up to 95% of the lateral elements paired in some nuclei. More of the *ph1b* nuclei showed complete pairing, suggesting that pairing was slower or less synchronized in the *Ph* nuclei. There was extensive pairing-partner switching in both genotypes (Figure 7), resulting in a maximum of 6 multivalents in one *Ph* nucleus and 11 multivalents in one *ph1b* nucleus.[225] The total length per nucleus of lateral elements involved in multivalent structures was several times greater in *ph1b* nuclei than in *Ph* nuclei,[225] supporting the idea that the *Ph* gene has an effect on the stringency of pairing.[222] However, the completeness of pairing possible in both genotypes suggests that the *Ph* gene effect on metaphase I pairing is probably manifested through control of crossing over in homoeologously paired SCs.[225] Since no attempt was made to substage the nuclei, Gillies[225] was not able to detect any evidence of pairing correction differences in the two genotypes. It is possible that most SCs formed in the *Ph* hybrid nuclei were between randomly paired nonhomologs, while the SCs in the *ph1b* nuclei were formed predominantly between homoeologous chromosomes. However, the presence in one zygotene *Ph* nucleus of a bouquet of 18 SC telomeres with matched ends[225] suggests that, at least at telomeres, pairing is not random.

The several studies of zygotene/pachytene SC formation in wheats have shown conclusively that there is pairing at this stage which produces partner switches and multivalents, and which most probably involves homoeologs.[218,220-223,225] Several studies revealed that such pairing occurs in both *Ph* genotypes and *ph* or nulli-5B genotypes; hence, there does not appear to be support for an intrinsic difference in the presynaptic arrangement of chromosomes in the two genotypes. Different doses of the *Ph* gene seem to have effects on the extent of pairing, the amount of multivalent pairing, and the correction of pairing.[221,222,225,229,230] There is also evidence that *Ph* suppresses crossing over between homoeologs.[222,225,229,230]

Although these effects seem disparate, they may be interrelated. Wischmann[221] has sug-

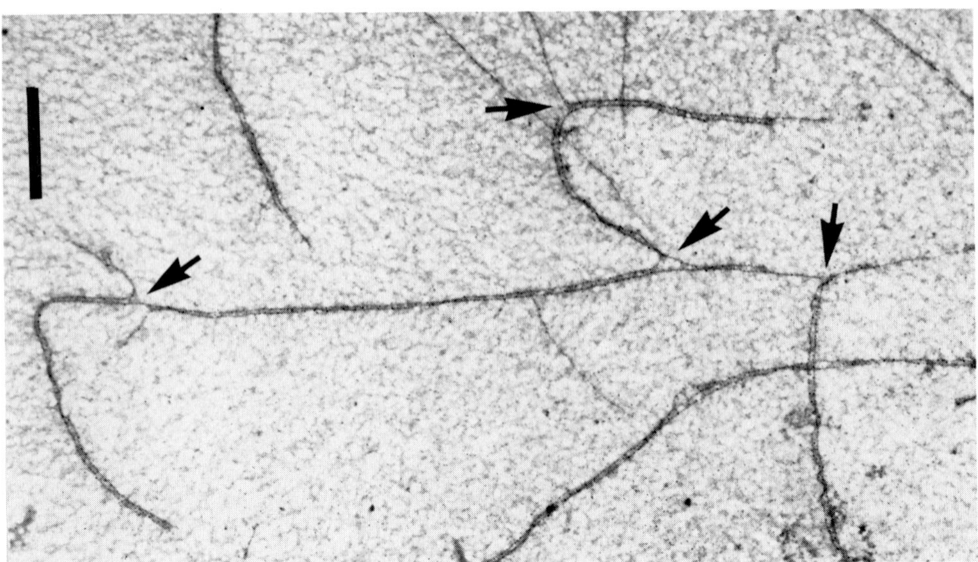

FIGURE 7.   Electron micrograph of a spread multivalent synaptonemal complex from a microsporocyte of pentaploid hybrid between *Triticum aestivum ph* and *T. kotschyi*, with pairing partner switch points arrowed. (Bar = 2 μm.) (From Gillies, C. B., *Theor. Appl. Genet.*, 74, 430, 1987. With permission.)

gested that persisting interlocks and incomplete pairing correction may be consequences of incomplete pairing. Toledo et al.[226] found that the lily hybrid "Black Beauty" (*Lilium speciosum* × *L. henryi*) had a mean of only 1.85 bivalents and 2.25 chiasmata per metaphase I cell, even though SCs were present at pachytene. Doubling the chromosome number to produce an amphidiploid resulted in apparently more complete pachytene pairing and a mean of 21.8 bivalents with 42.24 chiasmata at metaphase I. Biochemical analyses of the achiasmatic diploid hybrid[117] revealed that it had normal levels of the DNA unwinding (U) protein, but had little of the reannealing (R) protein which binds single-stranded DNA and, although there were normal amounts of endonuclease present, there was a low level of DNA nicking. DNA unwinding, nicking, and reannealing are presumed to be involved in crossing over.[227] The chiasmate allotetraploid had much increased nicking and appeared to have R protein levels similar to those in another chiasmate hybrid, leading Hotta et al.[117] to the conclusion that effective homologous pachytene pairing and R protein production were related. Since R protein reaches a maximum prior to completion of pairing, the authors concluded that it was the process of pairing which was important. Subsequently, it was found that a small nuclear RNA (P-snRNA) synthesized during homologous pairing at prophase I in lily, when added to the achiasmatic Black Beauty pachytene DNA, increased its accessibility to endonuclease, mimicking the behavior in the allotetraploid.[217] The conclusion was that, in lily, homologous or effective pairing regulates production of R protein which is necessary for crossing over. The *Ph* gene in wheat may have a similar regulatory effect,[222] perhaps by affecting the stringency of pairing necessary to trigger R protein production (see Reference 215).

V. CONCLUSIONS

Perhaps the main lesson to be drawn from the EM studies of SC formation in polyploids is that there is no clear delineation between autopolyploids and allopolyploids. As mentioned in the introduction to this paper, this is not unexpected. The major functional difference in pairing behavior between so-called autopolyploids and allopolyploids may be that allopolyploids possess a means of regulating pairing among multiple genomes. This allopolyploidization may reflect a genetic control of pairing, as epitomized by the *Ph* gene; it may be a consequence of the accumulation of chromosomal structural differences or the result of inherent differences in the genomes. Selection pressure for improved fertility will presumably act to favor and fix allopolyploid-like chromosome pairing behavior so that, with time, differences in pairing behavior become blurred.

Another interesting point to emerge is the prevalence of triple pairing (double SC), especially in situations where three or more copies of short chromosomes or segments are present at zygotene. Their occurrence challenges some ideas about the nature of SC formation, but there is no evidence to date that they have a significant role beyond early pachytene. Synaptic adjustment, multivalent correction, and elimination of triple pairing may all be manifestations of a process working to maximize the pairwise pachytene association of chromosomes, which normally will lead to the highest frequency of regular anaphase I disjunction and balanced fertile gametes.

ACKNOWLEDGMENTS

Without the help given by Malcolm Gillies, Sonya Gillies, Malcolm Ricketts, Stephen Edgar, and Ann Dollin, this Chapter would not have reached completion. I would also like to thank the many colleagues who have kept me informed of their work and generously provided illustrations.

REFERENCES

1. **Averett, J. E., Ed.**, Polyploidy in plant taxa: summary, in *Polyploidy, Biological Relevance*, Lewis, W. H., Ed., Plenum Press, New York, 1980, 269.
2. **Ohno, S., Ed.**, *Evolution by Gene Duplication*, Allen and Unwin, London, 1970.
3. **Swanson, C. P., Merz, T., and Young, W. J.**, *Cytogenetics. The Chromosome in Division, Inheritance and Evolution*, 2nd ed., Prentice-Hall, Englewood Cliffs, 1981.
4. **Hassold, T. J.**, Chromosome abnormalities in human reproductive wastage, *Trends Genet.*, 2, 105, 1986.
5. **Niebuhr, E.**, Triploidy in man. Cytogenetic and clinical aspects, *Humangenetik*, 21, 103, 1974.
6. **Mahony, M. J. and Robinson, E. S.**, Polyploidy in the Australian Leptodactyid frog genus *Neobatrachus*, *Chromosoma*, 81, 199, 1980.
7. **Bogart, J. P., Ed.**, Evolutionary significance of polyploidy in amphibians and reptiles, in *Polyploidy, Biological Relevance*, Lewis, W. H., Ed., Plenum Press, New York, 1980, 341.
8. **Rahn, I. M. and Martinez, A.**, Chromosome pairing in female and male diploid and polyploid anurans (Amphibia) from South America, *Can. J. Genet. Cytol.*, 25, 487, 1983.
9. **Schultz, R. J.**, Role of polyploidy in the evolution of fishes, in *Polyploidy. Biological Relevance*, Lewis, W. H., Ed., Plenum Press, New York, 1980, 313.
10. **Ferris, S. D. and Whitt, G. S.**, Phylogeny of tetraploid castomid fishes based on the loss of duplicate gene expression, *Syst. Zool.*, 27, 189, 1978.
11. **Uzzell, T.**, Meiotic mechanisms of naturally occurring unisexual vertebrates, *Am. Naturalist*, 104, 433, 1970.
12. **Goldstein, P. and Triantaphyllou, A. C.**, Karyotype analysis of the plant-parasitic nematode *Heterodera glycines* by electron microscopy. II. The tetraploid and an aneuploid hybrid, *J. Cell Sci.*, 43, 225, 1980.
13. **Goldstein, P. and Triantaphyllou, A. C.**, Pachytene karyotype analysis of tetraploid *Meloidogyne hapla* females by electron microscopy, *Chromosoma*, 84, 405, 1981.
14. **Muldal, S.**, The chromosomes of the earthworms. I. The evolution of polyploidy, *Heredity*, 6, 55, 1952.
15. **Lokki, J. and Saura, A.**, Polyploidy in insect evolution, in *Polyploidy. Biological Relevance*, Lewis, W. H., Ed., Plenum Press, New York, 1980, 277.
16. **Epstein, C. J.**, *The Consequences of Chromosome Imbalance. Principles, Mechanisms, and Models*, Cambridge University Press, Cambridge, 1986.
17. **Gillies, C. B. and Lesins, K.**, Pachytene chromosome morphology at different ploidy levels in *Medicago*, *Heredity*, 26, 486, 1971.

18. **Evans, G. M. and Davies, E. W.,** Fertility and stability of induced polyploids, in *Kew Chromosome Conference II,* Brandham, P. E. and Bennett, M. D., Eds., Allen and Unwin, London, 1983, 139.

19. **Stebbins, G. L., Ed.,** *Variation and Evolution in Plants,* Columbia University Press, New York, 1950.

20. **Burnham, C. R., Ed.,** *Discussions in Cytogenetics,* 5th Ed., C.R. Burnham Publishing, St. Paul, 1977.

21. **Moses, M. J.,** Synaptinemal complex, *Annu. Rev. Genet.,* 2, 363, 1968.

22. **Gillies, C. B.,** Synaptonemal complex and chromosome structure, *Annu. Rev. Genet.,* 9, 91, 1975.

23. **Lin, V. L.,** Fine structure of meiotic prophase chromosomes and modified synaptonemal complexes in diploid and triploid *Rhoeo spathacea, J. Cell Sci.,* 37, 69, 1979.

24. **Stack, S.,** The synaptonemal complex and the achiasmate condition, *J. Cell Sci.,* 13, 83, 1973.

25. **Moens, P. B.,** The structure and function of the synaptinemal complex in *Lilium longiflorum* sporocytes, *Chromosoma,* 23, 418, 1968.

26. **Moens, P. B.,** Synaptinemal complexes of *Lilium tigrinum* (triploid) sporocytes, *Can. J. Genet. Cytol.,* 10, 799, 1968.

27. **Moens, P. B.,** The fine structure of meiotic chromosome pairing in the triploid *Lilium tigrinum, J. Cell. Biol.,* 40, 273, 1969.

28. **Moens, P. B.,** The fine structure of meiotic chromosome pairing in natural and artificial *Lilium* polyploids, *J. Cell Sci.,* 7, 55, 1970.

29. **La Cour, L. F. and Wells, B.,** A study of the synaptonemal complex in water-treated pollen mother cells of *Phaedranassa viridiflora, Caryologia,* 27, 83, 1974.

30. **La Cour, L. F. and Wells, B.,** Deformed lateral elements in synaptonemal complexes of *Phaedranassa viridiflora, Chromosoma,* 41, 289, 1973.

31. **Loidl, J.,** SC-formation in some *Allium* species, and a discussion of the significance of SC-associated structures and of the mechanisms for presynaptic alignment, *Plant Syst. Evol.,* 158, 117, 1988.

32. **Rasmussen, S. W.,** Chromosome pairing in triploid females of *Bombyx mori* analyzed by three dimensional reconstructions of synaptonemal complexes, *Carlsberg Res. Commun.,* 42, 163, 1977.

33. **Rasmussen, S. W., Holm, P. B., Lu, B. C., Zickler, D., and Sage, J.,** Synaptonemal complex formation and distribution of recombination nodules in pachytene trivalents of triploid *Coprinus cinereus, Carlsberg Res. Commun.,* 46, 347, 1981.

34. **Loidl, J. and Jones, G. H.,** Synaptonemal complex spreading in *Allium.* I. Triploid *A. sphaerocephalon, Chromosoma,* 93, 420, 1986.

35. **Comings, D. E. and Okada, T. A.,** Triple chromosome pairing in triploid chickens, *Nature* (London), 231, 119, 1971.

36. **Sved, J. A.,** Telomere attachment of chromosomes. Some genetical and cytological consequences, *Genetics,* 53, 747, 1966.

37. **Von Wettstein, D., Rasmussen, S. W., and Holm, P. B.,** The synaptonemal complex in genetic segregation, *Annu. Rev. Genet.,* 18, 331, 1984.

38. **Therman, E. and Sarto, G. E.,** Premeiotic and early meiotic stages in the pollen mother cells of *Eremurus* and in human embryonic oocytes, *Hum. Genet.,* 35, 137, 1977.

39. **Hamey, Y., Abberton, M. T., Wallace, A. J., and Callow, R. S.,** Pairing autonomy and chromosome size, in *Kew Chromosome Conference III,* Her Majesty's Stationery Office, London, in press.

40. **John, B. and Henderson, S. A.,** Asynapsis and polyploidy in *Schistocerca paranenis, Chromosoma,* 13, 111, 1962.

41. **Rasmussen, S. W. and Holm, P. B.,** Chromosome pairing in autotetraploid *Bombyx* females. Mechanism for exclusive bivalent formation, *Carlsberg Res. Commun.,* 44, 101, 1979.

42. **Rasmussen, S. W.,** Chromosome pairing in autotetraploid *Bombyx* males. Inhibition of multivalent correction by crossing over, *Carlsberg Res. Commun.,* 52, 211, 1987.

43. **Wallace, B. M. N. and Hultén, M. A.,** Triple chromosome synapsis in oocytes from a human foetus with trisomy 21, *Ann. Human Genet.,* 47, 271, 1983.

44. **Speed, R. M.,** Meiotic configurations in female trisomy 21 fetuses, *Hum. Genet.,* 66, 176, 1984.

45. **Speed, R. M.,** Prophase pairing in a mosaic 18p-; iso 18q human female fetus studied by surface spreading, *Hum. Genet.,* 72, 256, 1986.

46. **Dollin, A. E. and Murray, J. D.,** Triple chromosome pairing in an aneuploid bull spermatocyte, *Can. J. Genet. Cytol.,* 26, 782, 1984.

47. **Stack, S.,** Two-dimensional spreads of synaptonemal complexes from Solanaceous plants. I. The technique, *Stain Technol.,* 57, 265, 1982.

48. **Gillies, C. B., Kuspira, J., and Bhambhani, R. N.,** Genetic and cytogenetic analyses of the A genome of *Triticum monococcum.* IV. Synaptonemal complex formation in autotetraploids, *Genome,* 29, 309, 1987.

49. **Byers, B. and Goetsch, L.,** Electron microscopic observations on the meiotic karyotype of diploid and tetraploid *Saccharomyces cerevisiae, Proc. Natl. Acad. Sci. U.S.A.,* 72, 5056, 1975.

50. **Borkhardt, B. and Olson, L. W.,** Meiotic prophase in diploid and tetraploid strains of *Allomyces macrogynus, Protoplasma,* 100, 323, 1979.

51. **Wagenvoort, M. and Ramanna, M. S.**, Identification of the trisomic series in diploid *Solanum tuberosum* L. group *Tuberosum*. II. Trivalent configurations at pachytene stage, *Euphytica*, 28, 633, 1979.

52. **Newton, W. C. F. and Darlington, C. D.**, Meiosis in polyploids. I. Triploid and pentaploid tulips, *J. Genet.*, 21, 1, 1929.

53. **Darlington, C. D.**, Meiosis in polyploids. II. Aneuploid hyacinths, *J. Genet.*, 21, 17, 1929.

54. **McClintock, B.**, The association of non-homologous parts of chromosomes in the mid-prophase of meiosis in *Zea mays*, *Z. Zellforsch. Mikrosk. Anat.*, 19, 191, 1933.

55. **Von Wettstein, D.**, The assembly of the synaptinemal complex, *Philos. Trans. R. Soc. London Ser. B*, 277, 235, 1977.

56. **Von Wettstein, D.**, The synaptinemal complex and four-strand crossing over, *Proc. Natl. Acad. Sci. U.S.A.*, 68, 851, 1971.

57. **Moses, M. J., Karatsis, P. A., and Hamilton, A. E.**, Synaptinemal complex analysis of heteromorphic trivalents in *Lemur* hybrids, *Chromosoma*, 70, 141, 1979.

58. **Gillies, C. B.**, The synaptonemal complex in higher plants, *CRC Crit. Rev. Plant Sci.*, 2, 81, 1984.

59. **Bojko, M.**, Presence of abnormal synaptinemal complexes in heterothallic species of *Neurospora*, *Genome*, 30, 697, 1988.

60. **Kuspira, J., Bhambhani, R. N., Sadasivaiah, R. S., and Hayden, D.**, Genetic and cytogenetic analyses of the A genome of *Triticum monococcum*. III. Cytology, breeding behaviour, fertility, and morphology of autotriploids, *Can. J. Genet. Cytol.*, 28, 867, 1986.

61. **Agarwal, P. K.**, Cytogenetical investigations in Musaceae. II. Meiotic studies in eight male sterile triploid banana varieties of India, *Cytologia*, 52, 451, 1987.

62. **Punyasingh, K.**, Chromosome numbers in crosses of diploid, triploid and tetraploid maize, *Genetics*, 32, 541, 1947.

63. **Darlington, C. D.**, Meiosis in diploid and tetraploid *Primula sinensis*, *J. Genet.*, 24, 65, 1931.

64. **Levan, A.**, Meiosis of *Allium porrum*, a tetraploid species with chiasma localisation, *Hereditas*, 26, 454, 1940.

65. **Gottschalk, W.**, Die Paarung homologer Bivalente und der Ablauf von Partnerwechseln in den frühen Stadien der Meiosis autopolyploider Pflanzen, *Z. Indukt. Abstamm. Vererbungs.*, 87, 1, 1955.

66. **Gottschalk, W.**, Polyploidy and its role in the volution of higher plants, in *Chromosome and Cell Genetics*, Sharma, A. K. and Sharma, A., Eds., Gordon and Breach, New York, 1985, 1.

67. **Gillies, C. B.**, Alfalfa chromosomes. II. Pachytene karyotype of a tetraploid *Medicago sativa* L., *Crop. Sci.*, 10, 172, 1970.

68. **Reddi, V. R.**, Pachytene pairing and the nature of polyploidy in *Sorghum arundinaceum*, *Caryologia*, 23, 295, 1970.

69. **Armstrong, K. C.**, Chromosome associations at pachytene and metaphase in *Medicago sativa*, *Can. J. Genet. Cytol.*, 13, 697, 1971.

70. **Beçak, M. L., Beçak, W., and Rabello, M. N.**, Further studies on polyploid amphibians *(Ceratophrydidae)*. I. Mitotic and meiotic aspects, *Chromosoma*, 22, 192, 1967.

71. **Rasmussen, S. W. and Holm, P. B.**, Mechanics of meiosis, *Hereditas*, 93, 187, 1980.

72. **Triantaphyllou, A. C. and Riggs, R. D.**, Polyploidy in an amphimictic population, *J. Nemat.*, 11, 371, 1979.

73. **Solari, A. J. and Moses, M. J.**, Synaptonemal complexes in a tetraploid mouse spermatocyte, *Exp. Cell. Res.*, 108, 464, 1977.

74. **Loidl, J.**, Synaptonemal complex spreading in *Allium*. II. Tetraploid *A. vineale*, *Can. J. Genet. Cytol.*, 28, 754, 1986.

75. **Loidl, J.**, Synaptonemal complex spreading in *Allium ursinum:* pericentric asynapsis and axial thickenings, *J. Cell Sci.*, 87, 439, 1987.

76. **Tschermak-Woess, E.**, Über chromosomale Plastizität bei Wildformen von *Allium carinatum* und anderen *Allium* - Arten aus den Ostalpen, *Chromosoma*, 3, 66, 1947.

77. **Kuspira, J., Bhambhani, R. N., and Shimada, T.**, Genetic and cytogenetic analyses of the A genome of *Triticum monococcum*. I. Cytology, breeding behaviour, fertility, and morphology of induced autotetraploids, *Can. J. Genet. Cytol.*, 27, 51, 1985.

78. **Morrison, J. W. and Rajhathy, T.**, Chromosome behaviour in autotetraploid cereals and grasses, *Chromosoma*, 11, 297, 1960.

79. **Rasmussen, S. W.**, Initiation of synapsis and interlocking of chromosomes during zygotene in *Bombyx* spermatocytes, *Carlsberg Res. Commun.*, 51, 401, 1986.

80. **Rasmussen, S. W. and Holm, P. B.**, The meiotic prophase in *Bombyx mori*, in *Insect Ultrastructure*, King, R. C. and Akai, H., Eds., Plenum Press, New York, 1982, 61.

81. **Sybenga, J.**, The quantitative analysis of chromosome pairing and chiasma formation based on the relative frequencies of MI configurations. VII. Autotetraploids, *Chromosoma*, 50, 211, 1975.

82. **Jackson, R. C. and Casey, J.**, Cytogenetic analyses of autopolyploids: models for triploids to octopoloids, *Am. J. Bot.*, 69, 487, 1982.

83. **Jackson, R. C. and Hauber, D. P.,** Autotriploid and autotetraploid cytogenetic analyses: correction coefficients for proposed binomial models, *Am. J. Bot.,* 69, 644, 1982.

84. **Jackson, R. C.,** Polyploidy and diploidy: new perspectives on chromosome pairing and its evolutionary implications, *Am. J. Bot.,* 69, 1512, 1982.

85. **Callow, R. S. and Galdwell, I.,** A general treatment of chromosome synapsis in even-numbered polyploids, *J. Theor. Biol.,* 106, 455, 1984.

86. **Callow, R. S. and Gladwell, I.,** Chromosome synapsis in odd-numbered polyploids, *J. Theor. Biol.,* in press.

87. **Gillies, C. B.,** An ultrastructural analysis of chromosome pairing in maize, *C.R. Trav. Lab. Carlsberg,* 40, 135, 1975.

88. **Holm, P. B.,** Three-dimensional reconstructions of chromosome pairing during the zygotene stage of meiosis in *Lilium longiflorum* (Thunb.), *Carlsberg Res. Commun.,* 42, 103, 1977.

89. **Hasenkampf, C. A.,** Synaptonemal complex formation in pollen mother cells of *Tradescantia, Chromosoma,* 90, 275, 1984.

90. **Gillies, C. B.,** An electron microscopic study of synaptonemal complex formation at zygotene in rye, *Chromosoma,* 92, 165, 1985.

91. **Timmis, J. N. and Rees, H.,** A pairing restriction at pachytene upon multivalent formation in autotetraploids, *Heredity,* 26, 269, 1971.

92. **Lentz, E. M., Crane, C. F., Sleper, D. A., and Loegering, W. Q.,** An assessment of preferential chromosome pairing at meiosis in *Dactylis glomerata, Can. J. Genet. Cytol.,* 25, 222, 1983.

93. **Murray, B. G., Sieber, V. K., and Jackson, R. C.,** Further evidence for the presence of meiotic pairing control genes in *Alopecurus* L. (Gramineae), *Genetica,* 63, 13, 1984.

94. **Lavania, U. C.,** High bivalent frequencies in artificial autotetraploids of *Hyoscyamus muticus* L., *Can. J. Genet. Cytol.,* 28, 7, 1986.

95. **Schifino, M. T. and Moraes Fernandes, M. I. B.,** Cytological evidence of genetic control, of chromosome pairing in artificial autotetraploids of *Trifolium riograndense* Burkart, *Genetica,* 74, 225, 1987.

96. **Avivi, L.,** The effect of genes controlling different degrees of homologous pairing on quadrivalent frequency in induced autotetraploid lines of *Triticum longissimum, Can. J. Genet. Cytol.,* 18, 357, 1976.

97. **Charpentier, A., Feldman, M., and Cauderon, Y.,** Genetic control of meiotic chromosome pairing in tetraploid *Agropyron elongatum.* I. Pattern of pairing in natural and induced tetraploids and in F1 triploid hybrids, *Can. J. Genet. Cytol.,* 28, 783, 1986.

98. **Simonsen, Ø.,** Cytogenetic investigations in diploid and autotetraploid populations of *Festuca pratensis, Hereditas,* 79, 73, 1975.

99. **Tai, W. and Dewey, D. R.,** Morphology, cytology, and fertility of diploid and colchicine-induced tetraploid crested wheatgrass, *Crop Sci.,* 6, 223, 1966.

100. **Ladizinsky, G.,** Genetic control of bivalent pairing in the *Avena strigosa* polyploid complex, *Chromosoma,* 42, 105, 1973.

101. **Avivi, L.,** Colchicine induced bivalent pairing of tetraploid microsporocytes in *Triticum longissimum* and *T. speltoides, Can. J. Genet. Cytol.,* 18, 731, 1976.

102. **Lelley, T., Mahmoud, A. A., and Lein, V.,** Genetics and cytology of unreduced gametes in cultivated rye (*Secale cereale* L.), *Genome,* 29, 635, 1987.

103. **Håkansson, A. and Levan, A.,** Endo-duplication meiosis in *Allium odorum, Hereditas,* 43, 179, 1957.

104. **Pijnacker, L. P. and Ferwerda, M. A.,** Development of the synaptonemal complex of two types of pachytene in oocytes and spermatocytes of *Carausius morosus* Br. (Phasmatodea), *Chromosoma,* 93, 281, 1986.

105. **Heslop-Harrison, J. S., Smith, J. B., and Bennett, M. D.,** The absence of the somatic association of centromeres of homologous chromosomes in grass mitotic metaphases, *Chromosoma,* 96, 119, 1988.

106. **Giraldez, R. and Santos, J. L.,** Cytological evidence for preferences of identical over homologous but not-identical meiotic pairing, *Chromosoma,* 82, 447, 1981.

107. **Santos, J. L., Orellana, J., and Giraldez, R.,** Pairing competition between identical and homologous chromosomes in rye and grasshoppers, *Genetics,* 104, 677, 1983.

108. **Naranjo, T. and Orellana, J.,** Meiotic behaviour of chromosomes 1R, 2R and 5R in autotetraploid rye, *Chromosoma,* 89, 143, 1984.

109. **Orellana, J. and Santos, J. L.,** Pairing competition between identical and homologous chromosomes in autotetraploid rye. I. Submetacentric chromosomes, *Genetics,* 111, 933, 1985.

110. **Benavente, E. and Orellana, J.,** Pairing competition between identical and homologous chromosomes in diploid and tetraploid cells of rye telotrisomic plants, *Can. J. Genet. Cytol.,* 28, 568, 1986.

111. **Sybenga, J.,** Preferential pairing in *Dactylis glomerata? Can. J. Genet. Cytol.,* 26, 640, 1984.

112. **Callow, R. S., Hamey, Y., and Pattrick, S. M.,** Pairing of identical chromosomes in an isogenic tetraploid, *Heredity,* 53, 107, 1984.

113. **Schlegel, R., Melz, G., and Nestrowicz, R.,** A universal reference karyotype in rye, *Secale cereale* L., *Theor. Appl. Genet.,* 74, 820, 1987.

114. **Snope, A. J.,** Meiotic behavior in autotetraploid maize with abnormal chromosome 10, *J. Hered.,* 58, 173, 1970.

115. **Macefield, A. J. and Evans, G. M.,** The effect of B chromosomes on meiosis in autotetraploid *Lolium perenne, Heredity,* 36, 393, 1976.

116. **Ingram, R. and Noltie, H. J.,** The control of chiasma frequency within a polyploid series in the genus *Senecio* (Compositae), *Genetica,* 72, 37, 1987.

117. **Hotta, Y., Bennett, M. D., Toledo, L. A., and Stern, H.,** Regulation of R protein and endonuclease activities in meiocytes by homologous chromosome pairing, *Chromosoma,* 72, 191, 1979.

118. **Watanabe, K.,** The control of diploid-like meiosis in polyploid taxa of *Chrysanthemum* (Compositae), *Jpn. J. Genet.,* 52, 125, 1977.

119. **Watanabe, K.,** Studies on the control of diploid-like meiosis in polyploid taxa of *Chrysanthemum,* I. Hexaploid *Ch. japonense* Nakai, *Cytologia,* 46, 459, 1981.

120. **Watanabe, K.,** Studies on the control of diploid-like meiosis in polyploid taxa of *Chrysanthemum.* II. Octoploid *Ch. jornatum* Hemsley, *Cytologia,* 46, 499, 1981.

121. **Watanabe, K.,** Studies on the control of diploid-like meiosis in polyploid taxa of *Chrysanthemum.* III. Decaploid *Ch. crassum* Kitamura, *Cytologia,* 46, 515, 1981.

122. **Watanabe, K.,** Studies on the control of diploid-like meiosis in polyploid taxa of *Chrysanthemum.* IV. Colchiploids and the process of cytogenetic diploidization, *Theor. Appl. Genet.,* 66, 9, 1983.

123. **Jackson, R. C. and Casey, J.,** Cytogenetics of polyploids, in *Polyploids. Biological Relevance,* Lewis, W. H., Ed., Plenum Press, New York, 1980, 17.

124. **Mulligan, G. A.,** Diploid and autotetraploid *Physaria vitulifera* (Cruciferae), *Can. J. Bot.,* 45, 183, 1967.

125. **Ellis, W. M., Lee, B. T. O., and Calder, D. M.,** Chromosome pairing in *Pao annua* L., *Can. J. Genet. Cytol.,* 15, 549, 1973.

126. **Harberd, D. J. and McArthur, E. D.,** Cyto-taxonomy of *Rhynchosinapis* and *Hutera* (Cruciferae — Brassiceae), *Heredity,* 28, 254, 1972.

127. **Sybenga, J.,** Allopolyploidization of autopolyploids. I. Possibilities and limitations, *Euphytica,* 18, 355, 1969.

128. **Sybenga, J.,** Allopolyploidization of autopolyploids. II. Manipulation of the chromosome pairing system, *Euphytica,* 22, 433, 1973.

129. **Doyle, G. G.,** The allotetraploidization of maize. IV. Cytological and genetic evidence indicative and substantial progress, *Theor. Appl. Genet.,* 71, 585, 1986.

130. **Gilles, A. and Randolph, L. F.,** Reduction of quadrivalent frequency in autotetraploid maize during a period of 10 years, *Am. J. Bot.,* 38, 12, 1951.

131. **Mastenbroek, I., DeWet, J. M. J., and Lu, C. -Y.,** Chromosome behaviour in early and advanced generations of tetraploid maize, *Caryologia,* 35, 463, 1982.

132. **Hazarika, M. H. and Rees, H.,** Genotypic control of chromosome behaviour in rye. X. Chromosome pairing and fertility in autotetraploids, *Heredity,* 22, 317, 1967.

133. **Crowley, J. G. and Rees, H.,** Fertility and selection in tetraploid *Lolium, Chromosoma,* 24, 300, 1968.

134. **Hossain, M. G.,** The significance of chromosome association in an advanced population of tetraploid rye, *Can. J. Genet. Cytol.,* 18, 601, 1976.

135. **Hossain, M. G. and Moore, K.,** Selection in tetraploid rye. I. Effects of selection on the relationships between seed-set, meiotic regularity and plant vigour, *Hereditas,* 81, 141, 1975.

136. **Hossain, M. G. and Moore, K.,** Selection in tetraploid rye. II. Effects of selection on the relationship between chiasma frequency and pairing configurations, *Hereditas,* 81, 153, 1975.

137. **Menzel, M. Y.,** Preferential chromosome pairing in allotetraploid *Lycopersicon esculentum — Solanum lycopersicoides, Genetics,* 50, 885, 1964.

138. **Menzel, M. Y. and Price, J. M.,** Fine structure of synapsed chromosomes in F1 *Lycopersicon esculentum — Solanum lycopersicoides* and its parents, *Am. J. Bot.,* 53, 1079, 1966.

139. **Attia, T., Busso, C., and Röbbelen, G.,** Digenomic triploids for an assessment of chromosome relationships in the cultivated diploid *Brassica* species, *Genome,* 29, 326, 1987.

140. **Schwarzacher-Robinson, T.,** Meiosis, SC-formation, and karyotype structure in diploid *Paeonia tenuifolia* and tetraploid *P. officinalis, Plant Syst. Evol.,* 154, 259, 1986.

141. **Sears, E. R.,** Genetic control of chromosome pairing in wheat, *Annu. Rev. Genet.,* 10, 31, 1976.

142. **Kerby, K. and Kuspira, J.,** The phylogeny of the polyploid wheats *Triticum aestivum* (bread wheat) and *Triticum turgidum* (macaroni wheat), *Genome,* 29, 722, 1987.

143. **Sears, E. R.,** Homoeologous chromosomes in *Triticum aestivum, Genetics,* 37, 624, 1952.

144. **Wall, A. M., Riley, R., and Chapman, V.,** Wheat mutants permitting homoeologous meiotic chromosome pairing, *Genet. Res.,* 18, 311, 1971.

145. **Riley, R. and Chapman, V.,** Genetic control of the cytologically diploid behaviour of hexaploid wheat, *Nature* (London), 182, 713, 1958.

146. **Wall, A. M., Riley, R., and Gale, M. D.,** The position of a locus on chromosome 5B of *Triticum aestivum* affecting homoeologous meiotic pairing, *Genet. Res.,* 18, 329, 1971.

147. **Riley, R.,** The diploidisation of polyploid wheat, *Heredity,* 15, 407, 1960.

148. **Riley, R. and Chapman, V.,** The effects of the deficiency of chromosome V (5B) of *Triticum aestivum* on the meiosis of synthetic amphiploids, *Heredity,* 18, 473, 1963.

149. **Riley, R.,** The basic and applied genetics of chromosome pairing, in *Proc. 3rd Int. Wheat Genetics Symp.,* Finlay, K. W. and Shepherd, K. W., Eds., Australian Academy of Science, Canberra, 1968, 185.

150. **Feldman, M.,** The effect of chromosomes 5B, 5D, and 5A on chromosome pairing in *Triticum aestivum, Proc. Natl. Acad. Sci. U.S.A.,* 55, 1447, 1966.

151. **Yacobi, Y. Z., Mello-Sampayo, T., and Feldman, M.,** Genetic induction of bivalent interlocking in common wheat, *Chromosoma,* 87, 165, 1982.

152. **Sears, E. R.,** An induced mutant with homoeologous pairing in common wheat, *Can. J. Genet. Cytol.,* 19, 585, 1977.

153. **Giorgi, B. and Barbera, F.,** Increase of homoeologous pairing in hybrids between a *ph* mutant of *T. turgidum* L. var. *durum* and two tetraploid species of *Aegilops: Aegilops kotschyi* and *Ae. cylindrica, Cereal Res. Commun.,* 9, 205, 1981.

154. **Jampates, R. and Dvořák, J.,** Location of the *Ph1* locus in the metaphase chromosome map and linkage map of the 5Bq arm of wheat, *Can. J. Genet. Cytol.,* 28, 511, 1986.

155. **Dvořák, J., Chen, K. -C., and Giorgi, B.,** The C-band pattern of a *Ph* mutant of durum wheat, *Can. J. Genet. Cytol.,* 26, 360, 1984.

156. **Mello-Sampayo, T.,** Genetic regulation of meiotic chromosome pairing by chromosome 3D of *Triticum aestivum, Nature New Biol.,* 230, 22, 1971.

157. **Sears, E. R.,** Mutations in wheat that raise the level of meiotic chromosome pairing, in *Proc. 16th Stadler Genetics Symp.,* Gustafson, J. P., Ed., Plenum Press, New York, 1984, 295.

158. **Ceoloni, C., Strauss, I., and Feldman, M.,** Effect of different doses of group-2 chromosomes on homoeologous pairing in intergeneric wheat hybrids, *Can. J. Genet. Cytol.,* 28, 240, 1986.

159. **Miller, T. E. and Reader, S. M.,** The effect of increased dosage of wheat chromosomes on chromosome pairing and an analysis of the chiasma frequencies of individual wheat bivalents, *Can. J. Genet. Cytol.,* 27, 421, 1985.

160. **Feldman, M. and Mello-Sampayo, T.,** Suppression of homoeologous pairing in hybrids of polyploid wheats X *Triticum speltoides, Can. J. Genet. Cytol.,* 9, 307, 1967.

161. **Hutchinson, J., Miller, T. E., and Reader, S. M.,** C-banding at meiosis as a means of assessing chromosome affinities in the *Triticeae, Can. J. Genet. Cytol.,* 25, 319, 1983.

162. **Charpentier, A., Feldman, M., and Cauderon, Y.,** Chromosomal pairing at meiosis of F1 hybrid and backcross derivatives of *Triticum aestivum* X hexaploid *Agropyron junceum, Can. J. Genet. Cytol.,* 28, 1, 1986.

163. **Dvořák, J.,** Chromosomal distribution of genes in diploid *Elytrigia elongata* that promote or suppress pairing of wheat homoeologous chromosomes, *Genome,* 29, 34, 1987.

164. **Chen, K. C. and Dvořák, J.,** The inheritance of genetic variation in *Triticum speltoides* affecting heterogenetic chromosome pairing in hybrids with *Triticum aestivum, Can. J. Genet. Cytol.,* 26, 279, 1984.

165. **Orellana, J., Cuñaado, N., and Cermeño, M. C.,** Genome-specific control of meiotic pairing evidenced in mutant *Aegilops ventricosa — Secale cereale* amphiploids, *Theor. Appl. Genet.,* 71, 532, 1985.

166. **Cermeño, M. C., Cuñado, N., and Orellana, J.,** Meiotic behaviour of Un, D and R genomes in the amphiploid *Aegilops ventricosa — Secale cereale* and the parental species, *Theor. Appl. Genet.,* 70, 679, 1985.

167. **Cuñado, N., Cermeño, M. C., and Orellana, J.,** Interactions between wheat, rye and *Aegilops ventricosa* chromosomes on homologous and homoeologous pairing, *Heredity,* 56, 219, 1986.

168. **Gupta, P. K. and Fedak, G.,** Genetic control of meiotic chromosome pairing in polyploids in the genus *Hordeum, Can. J. Genet. Cytol.,* 27, 515, 1985.

169. **Rajhathy, T. and Thomas, H.,** Genetic control of chromosome pairing in hexaploid oats, *Nature New Biol.,* 239, 217, 1972.

170. **Jauhar, P. P.,** Genetic regulation of diploid-like chromosome pairing in *Avena, Theor. Appl. Genet.,* 49, 287, 1977.

171. **Evans, G. M. and Aung, T.,** The influence of the genotype of *Lolium perenne* on homoeologous chromosome association in hexaploid *Festuca arundinacea, Heredity,* 56, 97, 1986.

172. **Evans, G. M.,** Genetic control of chromosome pairing in polyploids, in *Kew Chromosome Conference III,* Her Majesty's Stationery Office, London, in press.

173. **Lewis, E. J., Humphreys, M. W., and Caton, M. P.,** Disomic inheritance in *Festuca arundinacea* Schreb., *Pflanzenzüchtung,* 84, 335, 1980.

174. **Jauhar, P. P.,** Genetic control of diploid-like meiosis in hexaploid tall fescue, *Nature* (London), 254, 595, 1975.

175. **Evans, G. M. and Macefield, A. J.,** Suppression of homoeologous pairing by B chromosomes in a *Lolium* species hybrid, *Nature New Biol.,* 236, 110, 1972.

176. **Aung, T. and Evans, G. M.,** The potential for diploidizing *Lolium multiflorum* X L. *perenne* tetraploids, *Can. J. Genet. Cytol.,* 27, 506, 1985.

177. **Evans, G. M. and Davies, E. W.,** The genetics of meiotic chromosome pairing in *Lolium temulentum* X *Lolium perenne* tetraploids, *Theor. Appl. Genet.,* 71, 185, 1985.

178. **Jenkins, G.,** Synaptonemal complex formation in hybrids of *Lolium temulentum* X *Lolium perenne* (L.). II. Triploid, *Chromosoma,* 92, 387, 1985.

179. **Jenkins, G.,** Synaptonemal complex formation in hybrids of *Lolium temulentum* X *Lolium perenne* (L.). III. Tetraploid, *Chromosoma,* 93, 413, 1986.

180. **Jenkins, G. and Scanlon, M. J.,** Chromosome pairing in a *Lolium temulentum* X *Lolium perenne* diploid hybrid with a low chiasma frequency, *Theor. Appl. Genet.,* 73, 516, 1987.

181. **Miller, T. E.,** The homoeologous relationship between the chromosomes of rye and wheat. Current status, *Can. J. Genet. Cytol.,* 26, 578, 1984.

182. **Naranjo, T.,** Preferential occurrence of wheat-rye meiotic pairing between chromosomes of homoeologous group 1, *Theor. Appl. Genet.,* 63, 219, 1982.

183. **Bennett, M. D. and Kaltsikes, P. J.,** The duration of meiosis in a diploid rye, a tetraploid wheat and the hexaploid triticale derived from them, *Can. J. Genet. Cytol.,* 15, 671, 1973.

184. **Dhaliwal, H. S., Gill, B. S., and Waines, J. G.,** Analysis of induced homoeologous pairing in a *ph* mutant wheat X rye hybrid, *J. Hered.,* 68, 206, 1977.

185. **Jouve, N. and Giorgi, B.,** Analysis of induced homoeologous pairing in hybrids between 6x triticale *ph1* mutant and *Triticum aestivum* L., *Can. J. Genet. Cytol.,* 28, 696, 1986.

186. **Koebner, R. M. D. and Shepherd, K. W.,** Induction of recombination between rye chromosome 1RL and wheat chromosomes, *Theor. Appl. Genet.,* 71, 208, 1985.

187. **Koebner, R. M. D. and Shepherd, K. W.,** Controlled introgression to wheat of genes from rye chromosome arm 1RS by induction of allosyndesis. I. Isolation of recombinants, *Theor. Appl. Genet.,* 73, 197, 1986.

188. **Koebner, R. M. D. and Shephard, K. W.,** Controlled introgression to wheat of genes from rye chromosome arm 1RS by induction of allosyndesis. II. Characterisation of recombinants, *Theor. Appl. Genet.,* 73, 209, 1986.

189. **Alonso, L. C. and Kimber, G.,** The analysis of meiosis in hybrids. II. Triploid hybrids, *Can. J. Genet. Cytol.,* 23, 221, 1981.

190. **Kimber, G. and Alonso, L. C.,** The analysis of meiosis in hybrids. II. Tetraploid hybrids, *Can. J. Genet. Cytol.,* 23, 235, 1981.

191. **Lelley, T.,** Induction of homoeologous pairing in wheat by genes of rye suppressing chromosome 5B effect, *Can. J. Genet. Cytol.,* 18, 485, 1976.

192. **Dvorák, J.,** Effect of rye on homoeologous chromosome pairing in wheat X rye hybrids, *Can. J. Genet. Cytol.,* 19, 549, 1977.

193. **Schlegel, R. V. and Weryszko, E.,** Intergeneric chromosome pairing in different wheat-rye hybrids revealed by the giemsa banding technique and some implications on karyotype evolution in the genus *Secale, Biol. Zentralbl.,* 98, 399, 1979.

194. **Gupta, P. K. and Fedak, G.,** The inheritance of genetic variation in rye *(Secale cereale)* affecting homoeologous chromosome pairing in hybrids with bread wheat *(Triticum aestivum), Can. J. Genet. Cytol.,* 28, 844, 1986.

195. **Abirached-Darmency, M., Cauderon, Y., and Zickler, D.,** Meiotic chromosome pairing in three F1: *(Triticum — Secale)* hybrids. A comparative approach in light and electron microscopy, *Biol. Cell.,* 51, 365, 1984.

196. **Bennett, M. D., Dover, G. A., and Riley, R.,** Meiotic duration in wheat genotypes with or without homoeologous meiotic chromosome pairing, *Proc. R. Soc. London Ser. B,* 187, 191, 1974.

197. **Feldman, M.,** Regulation of somatic association and meiotic pairing in common wheat, in *Proc. 3rd Int. Wheat Genetics Symp.,* Finlay, K. W. and Shephard, K. W., Eds., Australian Academy of Science, Canberra, 1968, 169.

198. **Avivi, L. and Feldman, M.,** Arrangement of chromosomes in the interphase nucleus of plants, *Hum. Genet.,* 55, 281, 1980.

199. **Avivi, L., Feldman, M., and Brown, M.,** An ordered arrangement of chromosomes in the somatic nucleus of common wheat, *Triticum aestivum* L. II. Spatial relationships between chromosomes of different genomes, *Chromosoma,* 86, 17, 1982.

200. **Feldman, M. and Avivi, L.,** Ordered arrangement of chromosomes in wheat, in *Chromosomes Today,* Vol. 8, Bennett, M. D., Gropp, A., and Wolf, U., Eds., Allen and Unwin, London, 1984, 181.

201. **Yacobi, Y. Z., Levanony, H., and Feldman, M.,** An ordered arrangement of bivalents at first meiotic metaphase of wheat. I. Hexaploid wheat, *Chromosoma,* 91, 347, 1985.

202. **Yacobi, Y. Z., Levanony, H., and Feldman, M.,** An ordered arrangement of bivalents at first meiotic metaphase of wheat. II. Tetraploid wheat, *Chromosoma,* 91, 355, 1985.

203. **Dover, G. A. and Riley, R.,** The effect of spindle inhibitors applied before meiosis on meiotic chromosome pairing, *J. Cell Sci.,* 12, 143, 1973.

204. **Ceoloni, C., Avivi, L., and Feldman, M.,** Spindle sensitivity to colchicine of the *Ph1* mutant in common wheat, *Can. J. Genet. Cytol.*, 26, 111, 1984.
205. **Gualandi, G., Ceoloni, C., and Feldman, M.,** Spindle sensitivity to isopropyl-*N*-phenyl-carbamate and griseofulvin of common wheat plants carrying different doses of the *Ph1* gene, *Can. J. Genet. Cytol.*, 26, 119, 1984.
206. **Ceoloni, C. and Feldman, M.,** Effect of *Ph2* mutants promoting homoeologous pairing on spindle sensitivity to colchicine in common wheat, *Genome*, 29, 658, 1987.
207. **Shepard, J., Boothroyd, E. R., and Stern, H.,** The effect of colchicine on synapsis and chiasma formation in microsporocytes of *Lilium, Chromosoma*, 44, 423, 1974.
208. **Bennett, M. D., Toledo, L. A., and Stern, H.,** The effect of colchicine on meiosis in *Lilium speciosum* cv. Rosemede, *Chromosoma*, 72, 175, 1979.
209. **Hotta, Y. and Shepard, J.,** Biochemical aspects of colchicine action on meiotic cells, *Mol. Gen. Genet.*, 122, 243, 1973.
210. **Bennett, M. D., Stern, H., and Woodward, M.,** Chromatin attachment to nuclear membrane of wheat pollen mother cells, *Nature*, (London), 252, 395, 1974.
211. **Bennett, M. D., Smith, J. B., Simpson, S., and Wells, B.,** Intranuclear fibrillar material in cereal pollen mother cells, *Chromosoma*, 71, 289, 1979.
212. **Bennett, M. D. and Smith, J. B.,** The effect of colchicine on fibrillar material in wheat meiocytes, *J. Cell Sci.*, 38, 33, 1979.
213. **Driscoll, C. J., Bielig, L. M., and Darvey, N. L.,** An analysis of frequencies of chromosome configurations in wheat and wheat hybrids, *Genetics*, 91, 755, 1979.
214. **Driscoll, C. J., Gordon, G. H., and Kimber, G.,** Mathematics of chromosome pairing, *Genetics*, 95, 159, 1980.
215. **Driscoll, C. J.,** Mathematical comparison of homologous and homoeologous chromosome configurations and the mode of action of genes regulating pairing in wheat, *Genetics*, 92, 947, 1979.
216. **Catcheside, D. G.,** *The Genetics of Recombination*, Edward Arnold, London, 1977.
217. **Hotta, Y. and Stern, H.,** Small nuclear RNA molecules that regulate nuclease accessibility in specific chromatin regions of meiotic cells, *Cell*, 27, 309, 1981.
218. **Hobolth, P.,** Chromosome pairing in allohexaploid wheat var. Chinese Spring. Transformation of multivalents into bivalents, a mechanism for exclusive bivalent formation, *Carlsberg Res. Commun.*, 46, 129, 1981.
219. **Hobolth, P.,** Three-dimensional reconstructions of synaptonemal complexes at zygotene and pachytene in allohexaploid wheat var. Chinese Spring, in *Kew Chromosome Conference II*, Brandham, P. E. and Bennett, M. D., Eds., Allen and Unwin, London, 1983, 107.
220. **Jenkins, G.,** Chromosome pairing in *Triticum aestivum* cv. Chinese Spring, *Carlsberg Res. Commun.*, 48, 255, 1983.
221. **Wischmann, B.,** Chromosome pairing and chiasma formation in wheat plants triisosomic for the long arm of chromosome 5B, *Carlsberg Res. Commun.*, 51, 1, 1986.
222. **Holm, P. B., Wang, X. -Z., and Wischmann, B.,** An ultrastructural analysis of the effect of chromosome 5B on chromosome pairing in allohexaploid wheat, in *Kew Chromosome Conference III*, Her Majesty's Stationery Office, London, in press.
223. **Holm, P. B.,** Chromosome pairing and chiasma formation in allohexaploid wheat, *Triticum aestivum* analyzed by spreading meiotic nuclei, *Carlsberg Res. Commun.*, 51, 239, 1986.
224. **Gillies, C. B.,** The nature and extent of synaptonemal complex formation in haploid barley, *Chromosoma*, 48, 441, 1974.
225. **Gillies, C. B.,** The effect of *Ph* gene alleles on synaptonemal complex formation in *Triticum aestivum* × *T. kotschyi* hybrids, *Theor. Appl. Genet.*, 74, 430, 1987.
226. **Toledo, L. A., Bennett, M. D., and Stern, H.,** Cytological investigation of the effect of colchicine on meiosis in *Lilium* hybrid cv. ''Black Beauty'' microsporocytes, *Chromosoma*, 72, 157, 1979.
227. **Stern, H. and Hotta, Y.,** Regulatory mechanisms in meiotic crossing-over, *Annu. Rev. Plant Physiol.*, 29, 415, 1978.
228. **Loidl, J.,** The effect of colchicine on synaptonemal complex formation in *Allium ursinum*, *Exp. Cell Res.*, 178, 93, 1988.
229. **Wang, X.-Z. and Holm, P. B.,** Chromosome pairing and synaptonemal complex formation in wheat-rye hybrids, *Carlsberg Res. Commun.*, 53, 167, 1988.
230. **Holm, P. B. and Wang, X.-Z.,** The effect of chromosome 5B on synapsis and chiasma formation in wheat, *Triticum aestivum* cv. Chinese Spring, *Carlsberg Res. Commun.*, 53, 191, 1988.

INDEX

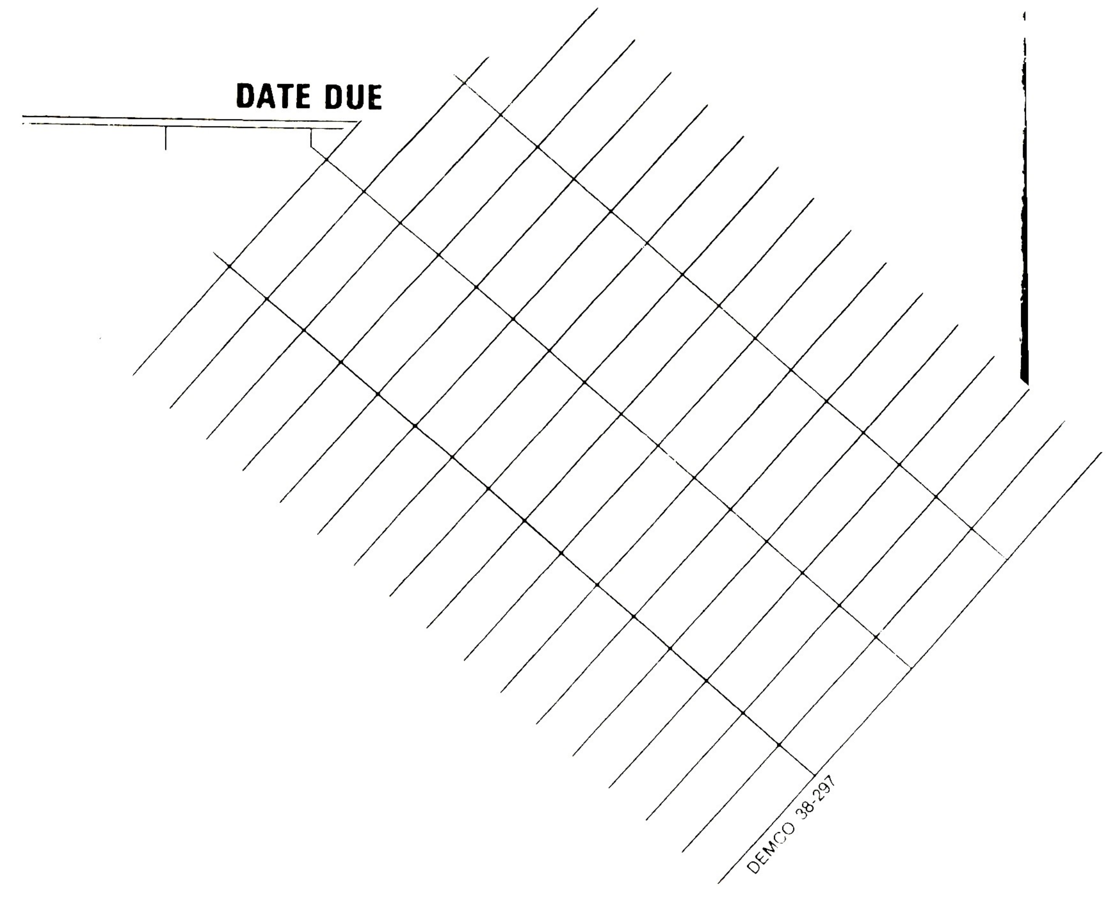

Fertility and chromosome
pairing